NUCLEAR POWER

Nuclear Accidents and Disasters

JAMES A. MAHAFFEY, PH.D.

An Infobase Learning Company

For Mary Carolyn Whatley

NUCLEAR ACCIDENTS AND DISASTERS

Copyright © 2012 by James A. Mahaffey, Ph.D.

Facts On File, Inc.
An imprint of Infobase Learning
132 West 31st Street
New York NY 10001

Library of Congress Cataloging-in-Publication Data
Mahaffey, James A.
Nuclear accidents and disasters / James A. Mahaffey.
p. cm.—(Nuclear power)
Includes bibliographical references and index.
ISBN 978-0-8160-7650-5
1. Nuclear accidents. 2. Nuclear reactor accidents. 3. Nuclear industry—Accidents. I. Title.
TK9152.M285 2011
363.17′99—dc22   2010046441

Facts On File books are available at special discounts when purchased in bulk quantities for businesses, associations, institutions, or sales promotions. Please call our Special Sales Department in New York at (212) 967-8800 or (800) 322-8755.

You can find Facts On File on the World Wide Web at http://www.infobaselearning.com

Text design by Annie O'Donnell
Composition by Adept Content Solutions
Illustrations by Bobbi McCutcheon
Photo research by Suzanne M. Tibor
Cover printed by Bang Printing, Brainerd, Minn.
Book printed and bound by Bang Printing, Brainerd, Minn.
Date printed: October 2011
Printed in the United States of America

10 9 8 7 6 5 4 3 2 1

This book is printed on acid-free paper.

Contents

Preface

Nuclear Power is a multivolume set that explores the inner workings, history, science, global politics, future hopes, triumphs, and disasters of an industry that was, in a sense, born backward. Nuclear technology may be unique among the great technical achievements, in that its greatest moments of discovery and advancement were kept hidden from all except those most closely involved in the complex and sophisticated experimental work related to it. The public first became aware of nuclear energy at the end of World War II, when the United States brought the hostilities in the Pacific to an abrupt end by destroying two Japanese cities with atomic weapons. This was a practical demonstration of a newly developed source of intensely concentrated power. To have wiped out two cities with only two bombs was unique in human experience. The entire world was stunned by the implications, and the specter of nuclear annihilation has not entirely subsided in the 60 years since Hiroshima and Nagasaki.

The introduction of nuclear power was unusual in that it began with specialized explosives rather than small demonstrations of electrical-generating plants, for example. In any similar industry, this new, intriguing source of potential power would have been developed in academic and then industrial laboratories, first as a series of theories, then incremental experiments, graduating to small-scale demonstrations, and, finally, with financial support from some forward-looking industrial firms, an advantageous, alternate form of energy production having an established place in the industrial world. This was not the case for the nuclear industry. The relevant theories required too much effort in an area that was too risky for the usual industrial investment, and the full engagement and commitment of governments was necessary, with military implications for all developments. The future, which could be accurately predicted to involve nuclear power, arrived too soon, before humankind was convinced that renewable energy was needed. After many thousands of years of burning things as fuel, it was a hard habit to shake. Nuclear technology was never developed with public participation, and the atmosphere of secrecy and danger surrounding it eventually led to distrust and distortion. The nuclear power industry exists today, benefiting civilization with a respectable percentage of the total energy supply, despite the unusual lack of understanding and general knowledge among people who tap into it.

This set is designed to address the problems of public perception of nuclear power and to instill interest and arouse curiosity for this branch of technology. *The History of Nuclear Power*, the first volume in the set, explains how a full understanding of matter and energy developed as science emerged and developed. It was only logical that eventually an atomic theory of matter would emerge, and from that a nuclear theory of atoms would be elucidated. Once matter was understood, it was discovered that it could be destroyed and converted directly into energy. From thre it was a downhill struggle to capture the energy and direct it to useful purposes.

Nuclear Accidents and Disasters, the second book in the set, concerns the long period of lessons learned in the emergent nuclear industry. It was a new way of doing things, and a great deal of learning by accident analysis was inevitable. These lessons were expensive but well learned, and the body of knowledge gained now results in one of the safest industries on Earth. *Radiation*, the third volume in the set, covers radiation, its long-term and short-term effects, and the ways that humankind is affected by and protected from it. One of the great public concerns about nuclear power is the collateral effect of radiation, and full knowledge of this will be essential for living in a world powered by nuclear means.

Nuclear Fission Reactors, the fourth book in this set, gives a detailed examination of a typical nuclear power plant of the type that now provides 20 percent of the electrical energy in the United States. *Fusion*, the fifth book, covers nuclear fission, the power source of the universe. Fusion is often overlooked in discussions of nuclear power, but it has great potential as a long-term source of electrical energy. *The Future of Nuclear Power*, the final book in the set, surveys all that is possible in the world of nuclear technology, from spaceflights beyond the solar system to power systems that have the potential to light the Earth after the Sun has burned out.

At the Georgia Institute of Technology, I earned a bachelor of science degree in physics, a master of science, and a doctorate in nuclear engineering. I remained there for more than 30 years, gaining experience in scientific and engineering research in many fields of technology, including nuclear power. Sitting at the control console of a nuclear reactor, I have cold-started the fission process many times, run the reactor at power, and shut it down. Once, I stood atop a reactor core. I also stood on the bottom core plate of a reactor in construction, and on occasion I watched the eerie blue glow at the heart of a reactor running at full power. I did some time in a radiation suit, waved the Geiger counter probe, and spent many days and nights counting neutrons. As a student of nuclear technology, I bring a near-complete view of this, from theories to daily operation of a power plant. Notes and apparatus from my nuclear fusion research have been requested by and given to the National Museum of American History of the Smithsonian Institution. My friends, superiors, and competitors for research funds were people who served on the USS *Nautilus* nuclear submarine, those who assembled the early atomic bombs, and those who were there when nuclear power was born. I knew to listen to their tales.

The Nuclear Power set is written for those who are facing a growing world population with fewer resources and an increasingly fragile environment. A deep understanding of physics, mathematics, or the specialized vocabulary of nuclear technology is not necessary to read the books in their series and grasp what is going on in this important branch of science. It is hoped that you can understand the problems, meet the challenges, and be ready for the future with the information in these books. Each volume in the set includes an index, a chronology of important events, and a glossary of scientific terms. A list of books and Internet resources for further information provides the young reader with additional means to investigate every topic, as the study of nuclear technology expands to touch every aspect of the technical world.

Acknowledgments

I wish to thank Dr. Douglas E. Wrege and Dr. Don S. Harmer, from whom I learned much as a student at the Georgia Institute of Technology in the schools of physics and nuclear engineering. They were kind enough to read the rough manuscript of this work, checking for technical accuracy and readability. Their combined wealth of knowledge in *nuclear physics* was essential for polishing this book. The manuscript also received a thorough cleansing by Randy Brich, a most knowledgeable retired USDOE health physicist from South Dakota, who is currently the media point-of-contact for Powertech Uranium. Special thanks to Kamara Sams of Environmental Communications, the Boeing Company, for providing important details and archived information concerning the Santa Susana Field Laboratory, Suzie Tibor for researching the photos, and Bobbi McCutcheon for the fine line art.

Introduction

The concept of deriving power from nuclear processes instead of from atomic processes was a sudden and exciting development in the mid-20th century. An example of an atomic process is the burning of coal, in which carbon is oxidized. It is a simple chemical reaction involving only the weak forces binding atoms together. An example of a nuclear process is *fission,* in which the powerfully bound *nucleus* of a *uranium* atom is blown asunder, releasing a burst of energy. A primary difference in these two processes is the magnitude of the energy involved per event, or the density of the energy. A nuclear process is at least a million times more energetic, with a million times more energy per reaction.

Advantage in the nuclear process was seen immediately upon discovery, with the volume of required fuel for a given energy product reduced by a factor of 1 million. Fuel would be so inexpensive as to be trivial, and there would be no atmospheric pollution produced by this new form of energy. In a world seeking better energy solutions, nuclear power seemed a positive development, and progress was made quickly in the 1940s and the 1950s.

Perhaps too much progress was made too quickly. Along with the dazzling, obvious advantages of nuclear power were new problems that had never plagued the power industry. Along with new ways to make steam were new ways to blow up a power plant. There was a greatly reduced atmospheric gases burden, but there were new, exotic types of waste to dispose of, dangerous in ways that had not been seen in the sweep of human experience. The new problems involved a class of invisible, undetectable rays, and one could be in jeopardy from proximity to nuclear power without even realizing it. The first use of nuclear processes had been, in fact, to wipe out two complete cities in Japan, ending World War II, and such power would require unusual handling. It had taken the industrial world generations to learn how to avoid burning down a factory with an errant candle flame or blowing up a building with a gas leak. Now it had a completely new, inherently strange set of lessons to learn.

Nuclear Accidents and Disasters is an examination of the learning process that has occurred over the last half century regarding the nuclear power industry. In chronological order, starting with the dim, first indications that there could be a problem, the narrative follows the many stages of this awakening. As the industry quickly moved from small, physically isolated experiments to a full-scale, applied power source, the importance of each accident or disaster grew, and the learning process became formalized. To read this account of the troubles that have plagued nuclear technology is to see the nuclear industry gradually mature. The reader will notice that the same mistake was seldom made three times. After seven chapters describing accidents of growing intensity, the narrative ends with an analysis of the massive *reactor* explosion at *Chernobyl* in the Ukraine region of the former Soviet Union.

A nuclear accident can involve an explosion, destroying equipment or an entire building and spreading *radioactive* material over a wide area. Several such accidents are described in this volume. The concept of an explosion has been established in the reader's mind by seeing it countless times on television and in movies. Think of an explosion, and the reader imagines a large, orange fireball and a great deal of yellow flame. In reality, that is not an accurate depiction of an explosion anywhere except in an oil refinery. Movie directors tend to enhance the drama of an explosion by including a few barrels of gasoline, so that there is a lot of color and a big ball of fire. In this book, the words "explosion" or "blast" usually refer to a steam detonation, in which a small volume of water suddenly becomes a large volume of steam. The results can be devastating, but there is no fireball. Even a hydrogen explosion, in which a volume of hydrogen oxidizes with severe force, makes no visible flame.

This volume is a concentrated depiction of the unique engineering problems that threatened to make nuclear technology impossible for society to accept. The solutions to these problems have changed industrial thinking and have modified the role of government control and regulation. *Nuclear Accidents and Disasters* describes the process through which nuclear power changed from a wild, seemingly haphazard novelty to a safety-conscious modern industry. Sidebars are interwoven throughout the text, introducing some interesting topics, such as President Jimmy Carter's experience with a reactor meltdown in Canada and the ghost village of Prypiat, Russia.

The accounts of accidents involving radiation have been written not to alarm the reader or to sensationalize accidents, but to give clear accounts of incidents that have been forgotten or never adequately publicized. Some events have only recently been declassified for public consumption, and this narrative also includes previously secret happenings in the former

Soviet Union. Nuclear power may be increasingly used as the world's energy demand expands, burnable fuel is used up, and atmospheric pollution levels become critical. It is important to become aware of the trials that have been experienced by the nuclear industry as an assurance that more and better control is now exercised and that the engineering has improved greatly. A glossary of terms useful for understanding the technical issues is included in the back matter, as well as a chronological list of the incidents described in the narrative and a list of current sources for further reading and study.

Expressing quantities of radiation when describing a nuclear accident can be difficult. Most accidents occurred in the 1950s or the 1960s, and radiation intensity and dose quantities in these accounts are expressed as originally measured or estimated. Doses are expressed in *roentgens, rads,* or *rems,* and the meaning of these measurements is not clear until many accidental doses and their consequences have been studied. The conversions among these units is not straightforward, but general trends become clear. Thousands of units of radiation dosage lead to death, hundreds lead to recoverable sickness, and tens of units are apparently harmless. These radiation dose units are now considered obsolete by the International System of Units (*SI,* abbreviated from the French Système International d'Unités), and where possible in this volume units are expressed both in the original form and in SI form. The obsolete units are still in regular use in the nuclear power industry, as most nuclear standards and documents were originally written using them and have not been converted.

Radiation intensity is expressed in the obsolete notation of *curies* and in the SI unit, the *becquerel.* One curie is roughly the activity of one gram of radium-226, which is a great deal of radiation, and one becquerel is one nuclear disintegration per second. The full meanings of these radiation, radiation dose, and radiation dose-rate units are explained in detail in another volume of this set, *Radiation,* but a complete understanding is not necessary to comprehend the ways that nuclear accidents and disasters have affected life on Earth.

1 New Problems in a New Industry

What makes accidents in the nuclear industry unique, interesting, and worthy of study are their potential for radiation release or contamination. Any other industrial process is certainly capable of causing great harm without warning. Steam boilers can blow up, poisonous gases can escape silently from a valve left open, and most industries use enormous loads of explosive or caustic chemicals, poised to become uncontrolled. There is no absolutely safe industry. A person can get killed in a cupcake factory, and one of the most dangerous explosives used in industry is powdered sugar. A lightning strike in the middle of an oil refinery can set a fire that burns for weeks, whose heat can be felt miles away.

As a threat to life, radiation occupies a special place in the list of human worries. Even a distant, unlikely threat of radiation exposure over the normal background radiation is a terror, and it alone stands in the way of nuclear power as a primary energy source for civilization. The danger of radiation has been known since before nuclear fission was discovered and nuclear power was developed, but still it took a great while for radiation exposure to become a fully recognized hazard with appropriate attention to its details. The maturing process for the nuclear industry has been long coming.

This chapter covers the pre-nuclear era, starting before the discovery of radiation, through the enlightened period of rapid discovery and industrialization of the newly found forces in nature. It is a micro-sample of the subsequent age of enhanced technology and applications of radiation on an enormous scale. In the early years, nothing was fully understood, and research and development ran completely out of phase with health concerns and control strategies. The dangers in those early days were tragically avoidable. Lessons were learned slowly and repeatedly. It was a small indication of the larger problem of dealing with megawatts of radiation production that would come as a by-product of nuclear power in the later 20th century. The sidebar "The Radithor Scandal" reveals details of one of the most extraordinary cases of death caused by a new industry.

THE DISCOVERY OF RADIATION SICKNESS

On a crisp autumn day in 1879 in the Ozark mountain country in Barry County, Missouri, three men were chasing down a wildcat, assisted by a hound. The dog was way ahead of the men, and he bounded through the woods and up a deep canyon. Bill Henry, John Dempsey, and Bill Boyceyer watched as the cat disappeared into a black hole in the side of the mountain. The dog, yelping enthusiastically, dived in after him. The three hunters were slightly more cautious than the dog, and they decided to wait at the entrance, guns cocked and ready for the cat to come blasting back out of the hole. They waited 15 minutes. The tormented feline seemed to have vanished. Finally the hound, still excited by the chase with his tail whipping back and forth, appeared at the top of the cliff, obviously having found that the hole ran clean through the mountain. This previously unknown cave was interesting, and it deserved some investigation.

The three entered the opening in the rock with due caution. Once inside they saw along the wall a vein of pure, silvery metal, and dollar signs came into their eyes. It was growing dark, and they decided to retire to the hunting camp and return the next day for a thorough exploration. The next morning they lit pitch-pine torches and proceeded inward. At about 200 feet in, the way was partially blocked by what appeared to be a large tree trunk of solid silver. It was the strangest metal they had ever seen, with the bluish sheen of a peacock's tail. In the yellow glare of the torches it seemed faceted, like a cut diamond. They chipped off a few pieces of the metal and put them in a small box. It was so heavy it was hard to carry.

They squeezed past the silvery metal and pressed on. About 500 feet from the entrance they entered an arched room. The walls of the room shone like polished silver, the floor was a light blue, and the ceiling was supported by three, transparent crystal columns. Excitement peaked. The men knew they had found their eternal fortunes. As they pressed past the

The entrance to a cave in Missouri, the site of the first documented case of radiation poisoning *(Charles Britton)*

columns, the torches started to die. Feeling overstimulated and a bit frightened, the three hunters made a quick retreat back to the cave entrance.

As they reached the mouth of the hole, Boyceyer's legs stopped working. He fell in a heap at the entrance, and shortly thereafter Henry keeled over in a faint. Dempsey started to go wild and talk crazy. The boy they had left sitting outside jumped up and ran down the mountain in near freefall, tearing into camp winded, trying to explain what had happened at the cave. A rescue team was quickly assembled.

Henry seemed the worse affected, and as his condition deteriorated they took him to a hospital in Carthage, Missouri, where he remained for several weeks in puzzling condition. Sores resembling burns broke out all over his body, and his legs seemed paralyzed. When he had recovered enough to walk, he staggered back to the cave to work his silver mine and reap his fortune, but a disagreement broke out with the person who claimed to own the land. Somewhat frustrated, he wound up blocking up the mouth of the cave and obscuring the entrance. The location of the cave faded from memories, and the story of the strange adventure of the three hunters became a spooky legend, to be told around campfires late at night in the hills of Arkansas and Missouri. Henry, Boyceyer, and Dempsey were probably the first human beings to suffer documented radiation poisoning.

In the formal physics lecture theaters and laboratories in Europe at the time, in the late 19th century, the concept of danger from invisible radiation was unknown. Academic scientists across the Continent and in the United Kingdom were busy studying the interesting properties of electricity in evacuated glass tubes. A standard piece of laboratory equipment was a Ruhmkorff coil, producing high-voltage electricity. All manner of effects were produced by electrifying tubes with these coils, and university research was progressing at a respectable pace. Working in parallel in the United States was an immigrant from Croatia, Nikola Tesla (1856–1943). Tesla had already made his reputation as a brilliant and productive inventor by creating the alternating current method of electrical power transmission for the Westinghouse Electric and Manufacturing Company. He now turned his attention to many disjointed projects, one of which was the production of high voltage. In Europe, they were working with tens of thousands of volts. Tesla worked with 4 million volts.

In April 1887, Tesla connected his extremely high voltage source to a single-electrode vacuum tube, having no anode, or return electrode. Electrons escaped the cathode in the tube running at extremely high speed. They crashed to a stop as they hit residual air particles in the imperfect vacuum, and Tesla noticed an "unusual radiation" resulting. Tesla liked to work with high-voltage with the lights turned off, and when he did so he noticed that the rays streaming out of his tube made the paint on the wall light up. Moreover, when he put up his hand to shadow the rays, they went right through it. He could see the bones in his hand, projected onto the wall. Tesla was fascinated, and he continued experiments with the invisible radiation.

Eight years later, on December 28, 1895, the discovery of the unusual radiation was formally announced, not by Tesla, but by Wilhelm Roentgen, working at the University of Munich. Like Tesla, he had noticed that the radiation passes through flesh but is stopped by bones or a metal object. In his paper in the *Proceedings of the Physical Medical Society*, Roentgen gave the phenomenon a temporary name: *X-rays*. Roentgen was cautious with his new, unknown rays. He covered his apparatus with lead, to direct the energy only forward.

Tesla had been casting about for an application of his X-rays. He had taken pictures of the bones in his hand, his knee, a bird, and a rabbit. He put his head in the beam and turned it up to full power. He felt a strange

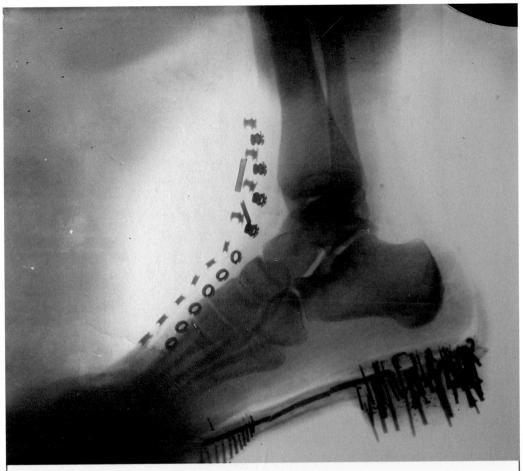

Nikola Tesla's X-ray of his foot, taken in 1895 (*Nikola Tesla Museum*)

sensation of warmth in the top of his head, shooting pains, and a shock effect in his eyes. He wrote three articles for the *Electrical Review* in 1896 describing what it felt like to be in an X-ray beam.

The effects were odd. "For instance," he first wrote, "I find there is a tendency to sleep and I find that time seems to pass quickly." He speculated that he had discovered an electrical sleep aid, safer than narcotics. In his next article for 1896, after having spent a lot of time being X-rayed, he observed "painful irritation of the skin, inflammation, and the appearance of blisters . . . and in some spots there were open wounds." In his final article of 1896, published on December 1, he advised staying away from X-rays, ". . . so it may not happen to somebody else. There are real dangers of Roentgen radiation."

This was the first mention in technical literature of the dangers of the newly discovered invisible rays. Tesla's equipment was powerful, and he was fortunate not to have died from his experiments. The dangers of the penetrating rays had been dimly perceived and noted.

In 1898, Marie and Pierre Curie discovered radium in their laboratory in Paris, France, and its invisible, energetic influences were formally named "radiation." Radiation had many curious properties, and one of them was to encourage the formations of sores on flesh that was exposed to it. The Curies dealt with tons of pitchblende and chalcolite uranium ore, refining them down to milligrams of dangerously radioactive radium. They tended to be careless in handling it. Marie enjoyed carrying a vial of a radium salt in a pocket of her lab coat because it glowed such a pretty blue color, and she would take it out and show visitors. Pierre would light a party at night using glass tubes, coated inside with zinc sulfide and filled with a radium solution. In the summer of 1903, the Curies were visited by Ernest Rutherford, a colleague working on the problem of atomic structure back in England. Rutherford could not help but notice Pierre's hands. They were swollen and terribly burned. He was in pain and seemed sickly. The exact nature of the emissions from radium had not been determined, but it was obvious to Rutherford that being too close to radium was not good for Pierre Curie.

In 1904, Thomas A. Edison (1847–1931), the "Wizard of Menlo Park" and the inventor of the electric lamp and the phonograph, had been experimenting with X-rays for several years. Edison thought of using X-rays to make a *fluorescent* lamp, and he proceeded to test a multitude of materials to find which one would glow the brightest under X-rays. His faithful assistant was a young, eager fellow, Charles M. Dally (1865–1904), who had worked for him for the past 14 years.

Dally was born in Woodbridge, New Jersey, and had served in the U.S. Navy for six years as a gunner's mate. After discharge from the navy, he signed on at the Edison Lamp Works in Harrison, New Jersey, as a glass blower, and in 1890 he moved to the Edison Laboratory in West Orange to work directly for Mr. Edison. Day after day, he held up screens of fluorescent material in front of an operating X-ray tube, staring directly at it to determine the quality of the light it produced. Nobody gave thought to any danger, but after a while Edison noticed that he could no longer focus his eye that he used to test a new fluoroscope, and "the X-ray had affected poisonously my assistant, Mr. Dally."

In the beginning, Dally's hair began to fall out and his face began to wrinkle. His eyelashes and eyebrows disappeared, and he developed a lesion on the back of his left hand. Dally usually held the fluorescent screen in his right hand in front of the X-ray tube and tested it by waving his left hand in the beam. There was no acute pain, only a soreness and numbness. Dally kept testing the fluorescent screens. He swapped hands, using his right to wave in front of the beam.

The lesion on his left hand would not heal, and conventional medical practice was at a loss to explain why. The pain became intolerable, and attempts to graft new skin onto the spreading sore were unsuccessful. The vascular system in the hand collapsed, and a cancer was detected at the base of the little finger. The hand was amputated at the wrist. Dally kept working on the X-ray project.

In the meantime, a deep ulceration developed on his right hand and four fingers had to be removed. Eventually, both arms had to be amputated, one at the shoulder and the other above the elbow. All efforts to stop the progression of the disease eventually failed, and Dally, after eight years of suffering, died in October 1904. Edison was shaken, and he dropped all work on the fluorescent lamp. "I am afraid of radium and polonium too," he commented, "and I don't want to monkey with them."

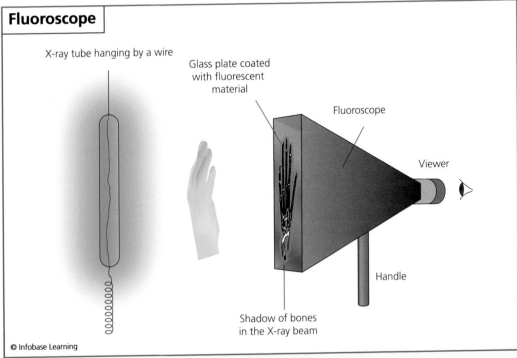

Fluoroscope

X-ray tube hanging by a wire

Glass plate coated with fluorescent material

Fluoroscope

Viewer

Handle

Shadow of bones in the X-ray beam

© Infobase Learning

Thomas Edison's assistant tested fluorescent materials by using a fluoroscope, repeatedly examining the shadow of the bones in his hand on the fluoroscope screen.

It took 11 years, but eventually news of the discovery of radium penetrated the Ozark hills, and in 1909 James L. Leib, a prospector and self-schooled geologist, saw a logical connection between the published properties of radium and the legend of the mysterious cave dating back to 1879. The spot price of radium at the time was, gram for gram, about 100 times the value of diamonds, or $70,000 per gram. Working quickly, Leib found the remaining member of the hunting party, Old Bill Boyceyer, still alive in Oklahoma. Boyceyer was glad to give what he could remember as directions to the hole, with a caution: Don't go in!

Leib found the cave, right where Boyceyer remembered, and he entered with unusual caution. He went in only far enough to pick up some bits of weird-looking, bluish rocks. Leib corresponded directly with Madame Curie, obtaining instructions for exposing photographic plates to the ore and confirming *radioactivity*. With the help of a photographer in Bentonville, Missouri, he succeeded. The few rocks he had brought back from the hole burned dark images into the plates, right through the dark slides and black paper wrapping. The radiographs were displayed at county fairs and apple shows all over the Ozarks, but Leib was unable to secure investors in a proposed radium mine.

In spring 1912, an enterprising man of vision from Chicago named John P. Nagel bought the land and commenced development as a mineral excavation site. Nagel proudly owned what was probably the world's most dangerous uranium mine. It was a confined space with no ventilation, and the stagnant air deep in the cave consisted primarily of radon gas. Workers in the cave would become dizzy breathing radioactive radon gas while caught in a crossfire of *alpha, beta,* and *gamma* radiation from all sides. The uranium and thorium and all their decay products, including radium, were radioactive.

Madame Marie Curie became head of a world center for the study of radioactivity, the Radium Institute, in Paris in 1918. She died on July 4, 1934, in a sanatorium in Geneva, Switzerland, of a blood disorder for which there was no cure. After many years of sickness, the disease was finally diagnosed as aplastic pernicious anemia. Her bone marrow, contaminated with radium, was unable to produce red blood cells, and the extensive exposure to X-rays during her medical volunteer work in World War I had contributed to the condition.

Her daughter, Irène Joliot-Curie, had taken up her mother's profession and also became a Nobel Prize–winning radiation scientist, working

THE RADITHOR SCANDAL

In 1918, Dr. William J. A. Bailey of the Bailey Radium Laboratories, Inc., of East Orange, New Jersey, began the manufacture of Radithor, an over-the-counter medicine advertised as "A Cure for the Living Dead." A bottle of Radithor, also called "Perpetual Sunshine," contained a dangerous dose of radium, dissolved in "triple distilled" water along with some thorium. It promised to cure arthritis, gout, neuritis, high blood pressure, mental retardation, and just about everything else by stimulating the endocrine system, but its money-making feature was a 17 percent rebate that a physician received on each prescription of Radithor.

Eben McBurney Byers, born in 1880, was a wealthy American socialite, athlete, and man-about-town who had inherited a steel conglomerate from his father, Alexander Byers. A graduate of Yale, Byers was chairman of the A. M. Byers Company of Philadelphia and the U.S. amateur golf champion of 1906. He had homes in three states, horse stables on two continents, and a room of trap-shooting trophies. On the return trip from the Harvard-Yale football game in 1927 on a chartered train, he was asleep when the car took a curve. He rolled out of the upper berth and hit the floor, hurting his arm.

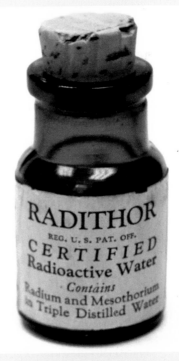

Byers complained to his physician, Dr. C. C. Moyar, about the persistent pain from the injury. Finding no evidence of bone breakage or anything that could be relieved by the usual means, the doctor prescribed Radithor. Byers drank a bottle of it after a meal, as specified, and he immediately felt better.

Reasoning that if a little Radithor is good then a lot of it will be marvelous, Byers proceeded to buy cases of it, at $30 per case, and drink it at least three times a day. He praised Radithor's healing properties to friends while

A bottle of Radithor, an over-the-counter medicine that proved deadly (*Oak Ridge Associated Universities*)

(continues)

(continued) _____

taking enormous doses of it, eventually downing 1,400 bottles and feeling better all the time. By 1930, he was not feeling better. He lacked that "toned-up feeling," lost weight, and suffered severe headaches. His jaw started to hurt and his teeth fell out. He stopped taking Radithor.

It was too late. He had more than 36 micrograms of radium deposited in his bones, over three times the lethal dose, and his health deteriorated in the most horrible way. The alpha radiation seemed to eat away his skeletal structure as the radium took the place of calcium in his bones. His long bones crumbled apart and lesions developed over his body. Hoping to save his life, surgeons removed all of his lower jaw and most of his upper jaw, with the exception of two front teeth. Holes began forming in his skull, and his brain abscessed. He died in agony on April 12, 1932.

The death of a wealthy, well-known figure by radiation poisoning received much publicity. The public sentiment was summed up in a headline in the *Wall Street Journal,* "The Radium Water Worked Fine Until His Jaw Came Off." The Radithor scandal created a heightened awareness of the dangers of radiation and radioactive materials; it brought an abrupt end to the radium-water industry; and it stirred the adoption of laws that would increase the powers of the U.S. Food and Drug Administration.

beside her in the Radium Institute. Joliot-Curie was working at her bench in the laboratory in 1946 when a sealed capsule of radioactive polonium exploded in her face. She contracted leukemia caused by her long-term exposure to radiation and the unfortunate large dose she received in the accident at the bench. She died on March 17, 1956, at the age of 58 in the Curie Hospital in Paris.

The Curies' famous laboratory was carefully restored and reconstructed in 1981 as a museum and as a monument to their pioneering work, but it was too radioactive for safe occupancy and had to be thoroughly decontaminated. Each speck of radium in the building, regardless of how small, had to be found, removed, and disposed of under strict rules and regulations. The world and its approach to radioactive contamination had improved since the Curies worked at the Radium Institute.

Nagel's radium mine closed down as the demand for the material waned, and the cave once again became a lost legend.

MASS CASUALTIES IN THE RADIUM INDUSTRY

It may seem unbelievable, but the industrial use of the extremely dangerous element radium continued in the United States at least through 1978. Radium contamination was still being cleaned up in the 1980s, and there is probably still more to be found. When mixed with a fluorescent powder, such as zinc sulfide, radium will glow with a greenish light, and it will do so for a long time, as the *half-life* for radium-226, the most common *isotope,* is 1,600 years. This is a highly desirable property of radium, as it can be used to light watch or alarm-clock dials, exit signs, instruments in airplanes, or even emergency landing lights at airports. The light requires no electricity, has no wires or moving parts, works under water, and is perpetually turned on. Many companies were formed in the early days, around World War I, and several fortunes were made using luminous, radium paint. It was sold under many brand names. The best was "Undark." The radium paint industry is an example of a lesson that was slow to be learned but may finally be under adequate control.

Beginning in 1917, radium-painted watch dial factories were established by the U.S. Radium Corporation in Orange, New Jersey, and then by Radium Dial Company in 1923 in Ottawa, Illinois. Soon there would be plants in Waterbury, Connecticut, and on Long Island, New York. Glow-in-the-dark watch dials were all the rage after World War I, and the companies prospered. With the introduction of the Westclox "Big Ben" alarm clock, business took off.

Young women were hired to paint the numbers on watch dials, making about $18 a week, which was excellent pay. Elsewhere, the same person would make $5 a week. An estimated 4,000 workers were hired to paint watches in the 1920s.

It required a bit of skill to fill in the tiny numerals on a watch dial with radium-loaded paint. The workers were trained to make a fine point on the brush by rolling the tip on the tongue or lips before it was dipped into the paint. "Wouldn't that get radium in my mouth?" a few would ask.

"Not to worry," said a supervisor. "If you swallow any radium, it'll make your cheeks rosy."

Not being complete fools, the supervisors, trainers, technicians, and chemists were extremely careful with handing the radium. They used lead screens, masks, and tongs and were careful not to get near the stuff, but the young women working on the watch-dial line were never told of any danger at all. The plant owners were in constant fear of labor unions, lawsuits, demands for hazard pay, or hysteria among the workers, so the workers were assured that the radium paint was harmless.

The job of painting watch dials could become tedious after a few units were finished, and the young women became bored. Just for fun, they would paint their nails with the glow-in-the-dark material, then go to the washroom and turn out the lights. Supervision became lax, and they started painting their teeth, their lips, and anything they wanted a boyfriend to find in the dark. It was only a matter of time before the radium collected in their bodies. It begins to replace the calcium in bones, and the high-energy alpha particles from the decaying radium are very destructive. The calcium replacement activity is apparently the most vigorous in the jaw and in the lower legs.

The workers began suffering from bone fractures, anemia from bone-marrow destruction, and necrosis of the jaw. Some grew large tumors on their jaws or leg bones. It was hard not to notice, but the plant owners admitted nothing.

One of the affected workers was a redhead named Margaret "Peg" Looney, a bookworm prone to an occasional giggling fit. She was hired at Radium Dial fresh out of high school in 1923. By 1929 her health was declining and, unknown to her, in her routine physical exam at the plant the doctors had detected radium in her system since 1925. She became too sick to work, and she had to leave the company on August 6, 1929. Eight days later, she died. She was 24 years old.

That was only the beginning, and back in New Jersey, at the U.S. Radium plant, worker Grace Fryer decided to sue for unsafe working conditions. It took two years to find a lawyer who would take the case, but four other workers joined the suit. The newspapers and the wire services picked up on the ensuing litigation, and the game was on. The five were named the "Radium Girls," and news from the courtroom circled the globe.

It was a difficult case to press, without precedent, but with irrefutable evidence of young women suffering from grotesque diseases. The results were significant. The right of an individual worker to sue for damages from corporations due to labor abuse was established. Scientific measurements of exhaled radon and radium excretions from dial painters were made, and for the first time, in 1941, the National Bureau of Standards established a baseline tolerance dose for radium uptake. As late as 1968, the trial still reverberated, as the Center for Human Radiobiology was established at the

Argonne National Laboratory in Illinois. The purpose of this center was to medically examine living dial painters and collect information. By 1993, 2,403 cases had been examined, with conclusions concerning the amount of radium uptake that a human body could tolerate. In 1982, the last of the dial-painting factories in Athens, Georgia, was decontaminated by the State of Georgia by excavating 18,000 cubic feet (540 m^3) of contaminated soil. Nearly 500 millicuries (1.85×10^{10} bq) of radium-226 were found and disposed of under tightly controlled conditions.

Each of the Radium Girls received $10,000 in the settlement.

2 Troubles at the Chalk River Nuclear Laboratories in Canada

At the start of the U.S. atomic bomb program, countries of the United Kingdom, allied in a common cause to stop the aggressive invasions of Germany and Japan, proved to be valuable resources. Both Canada and Great Britain loaned nuclear scientists from university positions to the research and development effort, and Canada was the only reliable source of uranium ore. Britain had concluded by 1940 that the construction of a uranium-based nuclear weapon was both possible and inevitable, but the entire United Kingdom lacked the necessary resources to pursue the matter.

After the end of the war, Chalk River, Canada, became a busy center of nuclear development in cooperation with renewed efforts in Great Britain. New, uniquely Canadian reactor technologies were produced. As was the case in every new nuclear research facility in the world, mistakes were made. The knowledge increased and matured at Chalk River, as the need for unusually vigilant safety procedures for nuclear materials and machines became clear. Lessons were learned.

This chapter examines three important accidents at Chalk River. In addition, a sidebar introduces a young U.S. Navy lieutenant, James Earl Carter, who participated in the after-incident cleanup of the most significant mistake at Chalk River, the meltdown of *NRX*. Lieutenant Carter would later become president of the United States and would be in office during a particularly serious nuclear disaster, the meltdown of the core in Unit 1 at the *Three Mile Island* Nuclear Generating Plant.

A PROBLEM WITH THE ZEEP REACTOR

In the late 1940s, the United States, the Soviet Union, Great Britain, and Canada started a variety of experimental reactor design programs. The programs varied widely, as the four groups of experiments diverged due to different approaches and enforced secrecy. A fundamental difference in reactor designs was the method of *neutron* moderation. The United States preferred an ordinary water or light water moderation design, assuming an *enriched uranium* fuel, or a non-moderated scheme, usually involving *plutonium* fuel and fast neutron fission. The British and the Soviets designed reactors using the traditional graphite *moderator* and natural uranium fuel. The Canadians carried on the original German reactor concept from World War II and built the first successful *heavy water* reactors.

In 1942, the British "Tube Alloys" atomic bomb project had moved to Montreal, Canada, and in late April 1944 British physicist Sir John Cockcroft came to Canada to lead the program. Experiments with lattice arrangements of uranium and heavy water, though never approaching *criticality*, indicated that such a reactor would be possible, and plans were made for a full nuclear reactor experiment, named "NRX," to be built in the remote Chalk River area. Cockcroft thought it a good idea, but first they should build a very low-power, heavy water moderated, experimental *pile*. In May 1944, he proposed the Zero Energy Experimental Pile. It was given the name "ZEEP." It would be built quickly, without impeding the slower work on the NRX. In August 1944, work began, with Lew Kowarski heading the project. Although there was pressure to rate the reactor at a power level of one kilowatt, it was successfully argued that it should be kept down to one watt. At higher power, much more shielding would be required, and the uranium and mechanical structures would become too radioactive to allow easy experimentation with different fuel lattice configurations.

Construction began at Chalk River on October 10, 1944, and was completed on September 4, 1945. The building looked like a small steel barn, and it was overwhelmed by the enormous brick NRX building being constructed next to it. The nuclear laboratory at Argonne, Illinois, provided some key materials and information from the operation of the CP-3 heavy water reactor, which had been operating with great success since May 15, 1944. ZEEP went critical at 3:45 P.M. on September 5, 1945. It was the first nuclear reactor in the world outside the United States.

The Zero Energy Experimental Pile, an early nuclear reactor experiment at Chalk River, Canada (*Canada Science and Technology Museum*)

ZEEP consisted of a cylindrical aluminum tank, mounted vertically in a large cube of graphite blocks, acting as a neutron reflector. Water tanks three feet (1 m) thick surrounded the graphite cube, to act as bioshielding. Fuel was loaded at the top, using an overhead crane, and boxes of borax-loaded wax were stacked on top of the reactor for removable bioshielding. Although there was no recirculating cooling system, there was a tank of heavy water in the basement of the building, and an electric pump was used to fill the reactor tank. The reactivity was controlled by varying the level of heavy water in the tank, using the pump and a dump valve connecting back to the basement tank. The control panel was located on the ground floor.

As a safety feature, the pump was automatically turned off after 10 seconds of running. This timer would prevent an unmonitored increase in reactivity by requiring an operator to sit at the controls and manually reset the pump switch over and over to increase the heavy water level in the reactor tank.

In 1946, and afterward from 1950 to 1955, ZEEP was in operation around the clock except for Sundays, when the reactor was shut down at 7:30 A.M. Sometime during the 1950–55 operating period, the reactor had been shut down for installation of neutron-detection foils. Two physicists were on top of the reactor, having removed the wax shielding and operating the overhead crane. The reactor tank had been drained, putting it in safe, subcritical condition. It took a long time to refill the reactor to operating level, so a technician at the control panel had already begun the process, resetting the pump switch every 10 seconds as the tank slowly filled.

The phone rang. Unfortunately, the phone was on the other side of the building. To keep the heavy water flowing while he answered the phone, the technician jammed the switch closed with a chip of wood. The pump kept going after the 10-second timed-out period. The technician was on the phone longer than he had expected.

ZEEP went supercritical with the two physicists standing on top with the bioshielding removed, as the heavy water level topped 52 inches (133 cm) in the tank. The reactor automatically scrammed (rapid emergency shutdown of a nuclear reactor) when the power level reached three watts, but it coasted upward by several factors of two as the heavy water drained out through the dump valve. The power excursion was large enough to scram the next-door operating NRX reactor through "skyshine" radiation, radiation scattered through the air over the ZEEP building and into the NRX building next door. The three men in the ZEEP building received significant radiation doses, but it is difficult to estimate the serious nature of the incident, as no one was wearing his dosimeter (device for measuring doses of radiation) at the time. As an industry rule, even in the early 1950s, everyone working near a nuclear reactor is required to wear a recording radiation counter to keep track of the total dose of radiation received, integrating over the period during which it is worn. Each instrument is read daily, and a log is kept of the total radiation received by each individual. The two physicists atop the reactor had left their dosimeters downstairs attached to their lab coats. The men were embarrassed by the folly of the accident, and it was never reported.

The ZEEP incident was notable on multiple levels. It showed that if a safety feature can be easily circumvented, then it is not a safety feature. There should be nothing to distract an operator and take him away from the control panel. Communications devices, such as telephones, should

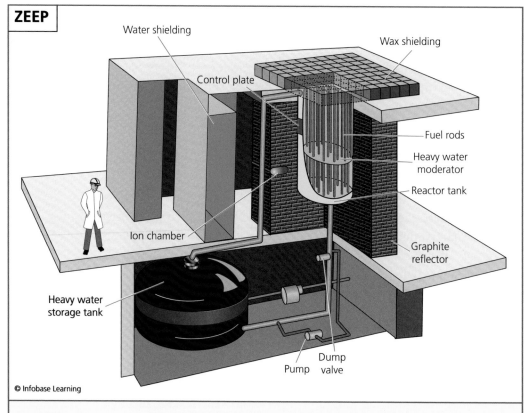

ZEEP

Water shielding

Wax shielding

Control plate

Fuel rods

Heavy water moderator

Reactor tank

Ion chamber

Graphite reflector

Heavy water storage tank

Dump
Pump valve

© Infobase Learning

The ZEEP was a small reactor, using heavy water as a moderator: The reactor was surrounded by graphite bricks and a water tank for radiation shielding.

be within easy reach of an operator. Whenever a researcher is working in a potentially hazardous situation, a health physicist should be on hand, doing nothing but monitoring the radiation and correct application of safety procedures. No one should be allowed to work in a radiation environment without an attached dosimeter. The scram mechanism for ZEEP was too sluggish for safe operation.

None of these points were factored into an improved understanding of research reactor safety, because the workers hid the incident from their superiors. The ZEEP accident was revealed decades later, as a history of the Chalk River reactors was being compiled. ZEEP otherwise had a long and productive life, and it was finally decommissioned in 1973. In 1997, it was dismantled and rebuilt in the Canada Science and Technology Museum in Ottawa, Canada, where it may be seen.

THE WORLD'S FIRST REACTOR MELTDOWN

The NRX at Chalk River was a heavy water–moderated, *light water*–cooled research reactor meant to test power reactor designs, make medical isotopes by neutron activation, and produce plutonium for the British atomic bomb development program. It was designed to run at 10 megawatts, and it was first brought to criticality in 1947. In 1954, it was upgraded to run at 42 megawatts, and at the time it was the most powerful research reactor in the world.

NRX was built as an aluminum cylinder, called the calandria, 9.8 feet (3 m) high and 26 feet (8 m) in diameter, mounted vertically. Into the calandria were placed 175 vertical calandria tubes, arranged in a hexagonal matrix. Each calandria tube was 2.4 inches (6 cm) in diameter, and it could hold a *control rod,* a specimen to be irradiated, or fuel. Most tubes held fuel, in aluminum-clad rods, each containing 121 pounds (55 kg) of natural uranium. Cooling water from the Ottawa River ran through the calandria tubes and around the fuel, at a rate of up to 66 gallons (250 l) per second. Control rods were made of boron carbide powder in steel tubes, with their insertion into the reactor controlled pneumatically.

The calandria was filled with 3,700 gallons (14,000 L) of expensive heavy water. Depending on the purity, that much heavy water would cost about $14 million. It was too expensive to use as coolant, so river water was used instead. The use of light water around the fuel degraded the moderating effect of the heavy water surrounding the calandria tubes, and if the light water were drained out or formed bubbles, the reactivity of the reactor would climb. To prevent corrosion of the metal parts in the calandria, helium gas was kept at constant pressure atop the heavy water. A separate helium tank with a movable top maintained the pressure. The changing height of the dome atop the helium tank indicated the flow of helium to and from the calandria.

On December 12, 1952, at 3:00 P.M., NRX was operating at 0.1 megawatts as researchers investigated the flow of coolant through the calandria tubes. Through a complicated series of operator errors and safety system failures, the NRX would be the first reactor in the world to experience a core meltdown. The first step in the accident began as a reactor operator in the basement mistakenly opened four valves. Although the control rods were normally operated by switches in the control room, there were air valves in the basement for emergency operation, for use only if the operating controls failed. Most of the handles had been taken off the valves so that they could not be actuated. An operator who happened to be in the

The NRX reactor at Chalk River, Canada, site of the world's first major nuclear reactor accident in 1952 (© Bettmann/CORBIS)

basement saw four valves with handles still attached, and he thought they were in the wrong position. He tried to set them right. Four control rods started moving out of the reactor, bringing it to supercriticality.

Red lights started coming on in the control room, indicating that the controls were moving. The supervisor immediately picked up the phone and dialed the basement. The operator picked up the phone, and he was sternly advised to stop opening the valves. Just to make sure that his wishes were carried out, the supervisor left the control room and ran down to the basement. He closed the valves and checked the air pressure.

The control rods should have fallen back into the calandria, and the red lights went out, indicating that they had. For mechanical reasons not understood, the control rods went down only enough to clear the switches for the red lights and then jammed. They were not all the way in, as they were supposed to be.

From the basement phone, the supervisor called his assistant in the control room, telling him to press the two buttons that would seal the pneumatic system and then to press the buttons that would push the control rods down, just to make sure everything was tight. All these control buttons were numbered, and, unfortunately, the supervisor remembered the numbers incorrectly. He gave the assistant the numbers for the buttons that withdrew an entire bank of control rods, instead of inserting them.

The assistant put down the phone and pressed the buttons, as instructed. The supervisor, realizing his error, started screaming into the phone, telling the assistant to stop. The assistant could not hear him, as the phone was not to his ear. The time was 3:07 P.M. With the control rods coming out, the power started rising quickly, doubling every two seconds. Noticing the unnatural rise in power, the assistant hit the big, red scram button.

The entire bank of controls should have slammed back into the reactor, but the erroneous actuation of buttons had not sealed the pneumatic system, as the supervisor had wished. Three rods stalled in the calandria tubes, and one was falling in very slowly. After 10 seconds, the power had risen to 17 megawatts, and the light water coolant began to boil. As this poor-quality component of the moderator developed bubbles and voids, the activity of the reactor suddenly improved, and the power rise increased sharply. Thinking quickly, someone in the control room hit the heavy water dump switch. Power continued to rise for another five seconds, reaching 80 megawatts. With the heavy water level in the calandria falling quickly, the reactor power dropped to zero 25 seconds later.

The accident was not over. The dome on the helium tank was falling rapidly. The assistant thought about what this meant. The tank and its helium supply were sealed. The heavy water was running out the bottom so fast that there was danger of imploding the calandria from the vacuum being produced, so he slammed the dump valve closed. The helium tank dome continued to drop, and it eventually hit its lowest possible point and jammed. This could only mean that the calandria tubes had burst, and the helium was escaping. With the helium supply exhausted, air rushed in through the breached tubes, hit the hydrogen bubbles that had formed in the hot fuel, and exploded. At 3:11 P.M., the top of the helium tank was suddenly blown forcibly to its highest possible level. At 3:17 P.M., an automatic air-quality siren went off, indicating that everyone at Chalk River should stay indoors. One fuel rod in NRX had been experimentally air-cooled, and its failure had put *fission products* from 66 pounds (30 kg) of uranium up the vent stack on the building. A plume of radioactive smoke 2,600 feet (800 m) wide carried the pulverized debris downwind. Staff in a neighboring building called the NRX control room and wanted to know what was going on. Their radiation monitors had gone off scale.

The reactor was finally secured in the subcritical condition, but the residual heat from the 80 megawatt excursion had to be removed, so the river water system was not shut down. With the calandria tubes broken, water leaked out of the reactor and onto the floor of the reactor building. By 6:00 P.M., the floor was awash. In a few days, the water level reached three feet (1 m), and it spilled into the helium tank room and the two heavy water storage rooms. In all, 1 million gallons (4.5 million l) of water contaminated with a wide spectrum of radioactive fission products filled the basement of the NRX building. Fuel in the NRX, deprived of contact with cooling water while undergoing an extremely high power excursion, had melted and broken through the aluminum calandria tubes. The fuel matrix and the calandria were beyond repair. The entire reactor tank, its internal structure, and the fuel would have to be replaced, and the wrecked equipment would have to be buried.

It took close to two years to clean up the radiation spills at NRX and rebuild the calandria and the reactor core. The NRX was restored to operation, and it was used for nuclear research and isotope production for 45 years. It was shut down permanently in 1992. Lessons learned in the 1952 accident shook the entire world of nuclear research and have affected all reactor system designs since. The weakness of the NRX shutdown

FORMER PRESIDENT JIMMY CARTER (1924–   ): A FORMATIVE EXPERIENCE

James Earl "Jimmy" Carter was born in the small, agricultural community of Plains, Georgia, on October 1, 1924. His father, James, was a prominent businessman in the community, and his mother, Lillian, was a registered nurse. The Carter family owned a large peanut farm in Plains, and while in high school Jimmy Carter became a fully participating member of the Future Farmers of America. He was the eldest of four children.

After high school, he attended Georgia Tech and Georgia Southwestern State University before his appointment to the U.S. Naval Academy in 1943. His class of 1947 graduated in 1946 under an accelerated schedule due to World War II, and Carter was ranked 59th out of a class of 820. His degree was a bachelor of science in physics. He was commissioned ensign, USN, in June 1946, and his first duty was aboard the USS *Wyoming* in August 1946. In June 1948, Carter took instruction at the USN submarine school in New London, Connecticut, and he saw duty aboard the diesel-electric submarine USS *Pomfret*.

The USS *Barracuda* was a new snorkel-equipped diesel sub, built in 1949, having an experimental sonar array, the BQR-4, mounted on the bow. Carter won assignment to this advanced submarine, and he was promoted to lieutenant in June 1952. In October of the same year, he gained duty with the *Atomic Energy Commission (AEC)*, for assisting in the design and development of nuclear propulsion plants for naval vessels. Before the assignment, he was subjected to an interview with the head of all nuclear operations, the famous admiral Hyman G. Rickover, a man well known for his soul-crushing interrogations. When Carter under questioning admitted that perhaps he had not consistently done his best at the Academy, Rickover asked a simple, pointed question: "Why?"

Rickover in 1952 was in the middle of his intense quest to build the first nuclear-powered submarine, and he was using every resource he could gather. He asked for and was granted permission to test some submarine reactor fuel assemblies in the NRX reactor at Chalk River. In December, two months after Carter had entered the nuclear program, NRX was suddenly out of commission, with a wrecked reactor tank and contaminated water flooding the building. Rickover was quick to offer assistance. He sent 150 navy personnel, all with at least an inkling of nuclear experience

(continues)

(continued) _____

or some training, to Chalk River, with Lieutenant Jimmy Carter in command. The Canadian military supplied 170 personnel, 20 construction workers were employed, and the staff of 862 at Chalk River all helped in the cleanup.

The first problem was disposing of the water in the basement, and this was handled by building a pipeline to a sandy area one mile (1.6 km) away. The water was allowed to soak into the sand and seep away, leaving about 10,000 curies of long-lived fission products trapped in the soil. Without the water problem, the reactor building was left with insoluble, radioactive residue contaminating the floor and walls. By the radiation standards of the time, an entire year's exposure limit was 15 rem (0.15 Sv) per person. At the rate of radiation exposure in the basement, a person could receive an entire year's dosage in a few hours. Each worker could scrub the basement only for a very short time.

His experience with the NRX cleanup scarred Carter for life, and it affected his perceptions and attitudes toward civilian nuclear power production. Carter was to be assigned duty aboard the experimental nuclear submarine USS *Seawolf*, then under construction, but his father died in July 1953. He resigned his commission abruptly and returned to Plains, Georgia, to run the family's peanut farm.

Discovering an interest in politics, Carter started with a position on the local library board, jumped to the Georgia State Senate, and went on to become the governor of Georgia. His next move was a leap, as he became president of the United States in 1976. He ran a campaign promising to shut down the civilian nuclear power industry, among other issues. On April 7, 1977, President Carter moved to deny an operating license to the new nuclear *fuel reprocessing* plant at Barnwell, South Carolina. With the veto of S. 1811, the ERDA Authorization Act of 1978, he stopped federal funding on the Clinch River liquid sodium–cooled fast *breeder reactor* in Tennessee. At the beginning of 1979, the energy market exploded, with gasoline prices rising rapidly, and on March 28, 1979, reactor Unit 2 at the Three Mile Island Nuclear Generating Station near Harrisburg, Pennsylvania, experienced a disaster. Equipment failures and operators scrambling to compensate resulted in a meltdown of the reactor core and a total loss of the power unit.

Jimmy Carter was the first president to have a degree in physics, the first to have experience with nuclear technology, and the first to lose a 786-megawatt power station due to a fuel melt. This accident would affect the structure and the future planning of the electrical power industry in the United States for decades.

system was particularly evident, as it could easily be disabled simply by opening valves or pushing buttons. All scram systems designed since the NRX incident cannot be tampered with, either accidentally, purposefully, or by mechanical or electrical failure. Shutdown systems are now built so that any misplacement of a switch or a valve will cause a shutdown and not prevent it. The efficiency, simplicity, and direct nature of the man-machine interface in a nuclear reactor have been constantly improved, to minimize the possibility of wrongly actuating a control or misinterpreting instructions. No switch or button in a reactor control room is designated by a simple number. Each actuation point on a control console is labeled by its function.

Still, there remains a human factor that is persistent in nuclear operations and in general industry. When operators see a problem arising, either through error or uncontrollable influences, they will try to correct it. Decisions must be made in haste, as a problem develops quickly. Sometimes, operator action only makes the problem worse. The NRX incident would not be the last core meltdown.

A FIRE IN THE NATIONAL RESEARCH UNIVERSAL REACTOR

Chalk River Laboratories in Ontario, Canada, remains a major center of nuclear research and development, with ongoing reactor operations. The ZEEP reactor was shut down in 1973, and the NRX ceased operations in 1992. The 10-kilowatt Pool Test Reactor, or PTR, ran from 1957 until 1990. SLOWPOKE, the Safe Low-Power Critical Experiment, was built in 1970, but in 1971 it was moved to the University of Toronto. MAPLE-1 and MAPLE-2, the Multipurpose Applied Physics Lattice Experimental reactors brought to power in 2000 and 2003, are still running, as is ZED-2, the Zero Energy *Deuterium* reactor built in 1960.

The crown jewel at Chalk River is the *NRU*, or the National Research Universal reactor. Encouraged by success with the NRX design, in 1949 the National Research Council of Canada began design on an even larger, more versatile reactor, suitable for both deuterium-moderated power reactor experiments and for isotope production. Similar to the NRX, the NRU was designed with an aluminum calandria and would burn natural, metallic uranium in vertically mounted aluminum tubes. It was initially to run at 200 megawatts, 10 times as powerful as NRX. Eight years passed from inception to first operation. NRU went critical on November 3, 1957.

The NRU national research universal reactor at Chalk River, Canada, the oldest nuclear reactor still in service (© Bettmann/CORBIS)

There are two problems with using natural, metallic uranium fuel. Natural fuel in a high-power reactor must be changed out frequently. To avoid lengthy shutdowns, much design work has gone into automated or remotely controlled fuel handler systems for natural-fuel reactors, so that fuel can be replaced in the reactor core quickly and without subjecting workers to unnecessary exposure to radiation from fission products in the expended fuel. Enriched uranium can provide power for years in a reactor core, and in that case it is easier just to dismantle the reactor and change out the fuel, even if it means a lengthy down period. The automated fuel-handler systems can be rough on the uranium assemblies, not giving them the careful, individual attention of a human-driven refueling crew.

Using metallic fuel is also problematic. Enriched fuel is usually in an oxide or carbide form. It is a nonburnable ceramic material, with a high melting temperature and no ability to combine further with oxygen.

Metallic uranium can catch fire and burn in air. To prevent this, it is usually sealed in a thin, aluminum can. It is fragile, and if the aluminum can is broken, the uranium is exposed to air. For natural uranium, all advantage must be given it to achieve criticality, and this means that the maximum density of pure metal is necessary.

In the first year of operation of the NRU, on Friday, May 23, 1958, the NRU had been running steadily for a week, but for unknown reasons the power began to rise and the reactor automatically scrammed to full shutdown condition. Having no explanation for the shutdown, the operating crew tried to restart the reactor. It scrammed again, this time setting off radiation alarms. A quick analysis of the alarm conditions showed that three fuel assemblies had broken open and were spilling fission products. The worst case was the fuelrod in channel J-18. With the reactor in full shutdown condition, the robotic fuel-handling crane was put into action to remove the damaged units, starting with J-18.

The fuel-handling machine was an elaborate device, two stories tall and moving on rails above the reactor. A large flask of heavy water was used to immerse a fuel assembly in coolant as it was plucked from the top of the reactor, to prevent it from bursting into flame, with a long, metal snout being lowered into the core at the fuel channel location. By late in the evening on May 24, the operating crew had positioned the fuel handler over J-18, and the machine automatically grabbed the fuel and raised it into the snout. Unknown to the operators, a valve had jammed open, and all of the cooling heavy water had drained out of the flask. There was nothing to keep the damaged fuel from burning. Seeing a problem developing, the operator tried to put the fuel back in the reactor core. It was warped and damaged, and it jammed in the fuel channel. He pulled it back out and tried to move the handler across the building, where he could drop the fuel into a waiting pool of water. The handler refused to move. Interlock switches prevented the machine from being operated if it detected no heavy water in the snout. The situation was becoming critical, as the fuel had been without cooling water for 12 minutes. Operators rushed to electrically bypass the safety interlocks on the fuel handler, so that the machine could transport the fuel. The uranium caught fire, and the radiation level in the building started to reach dangerous levels.

Burning fuel is serious, as it quickly sprays radioactive fission products into the air as smoke. As it jerked the fuel out of the core, the automatic machine had ripped it in two. The larger length of fuel was in the robotic gripper, but as it was trying to pass it overhead to the cooling pool, a one-foot (0.3-m) section of the fiercely burning uranium broke off and hit the floor in a shallow maintenance pit. The building ventilation system was jammed in the open position, and smoke pouring out of the burning fuel rapidly contaminated the entire building and a sizeable area downwind from NRU.

A relay team of all available scientists, technicians, accountants, and bookkeepers was assembled quickly to work on the fire. They donned protective gear, meant to keep the radioactive smoke out of their lungs and off their skin. Each took one full-speed running pass at the fire, into the building, up a long, precarious steel stairway, dumping a bucket of wet sand on the flaming fuel assembly. In 15 minutes, the fire was out, but it would take three months to clean up all the spilled fuel fragments and residual contamination, involving more than 600 men. In August 1958, the NRU resumed operation, and it has been in use ever since.

The power level of the NRU was downgraded to 60 megawatts, and it was converted to using highly enriched uranium fuel in 1964. In 1991, it was further modified to run at 135 megawatts using a lower enrichment percentage of U-235 in the fuel. There was no official medical follow-up to see whether the workers involved in the building cleanup following the fire ever exhibited higher-than-normal incidences of cancer, because care had been taken to ensure that no one person received more than a permissible dose of radiation.

The NRU reactor is still in heavy use, and it is the world's largest medical isotope production facility, supplying the Western Hemisphere its only source of diagnostic and therapeutic radioactive materials. Each year *cobalt-60* produced in the NRU is used for cancer therapy on 15 million patients in 80 countries. The production of technetium-99m alone accounts for 80 percent of the nuclear pharmaceuticals used in the United States and is used in CAT and SPECT scans for diagnosis for more than 5 million patients every year. NRU also produced large quantities of xenon-135, iodine-131, and iodine-125, which are used in a variety of diagnostic and disease treatment applications. Carbon-14 produced at NRU is sold to chemistry, bioscience, and environmental labs for use as a tracer component in biochemicals. Iridium-192 from NRU is used as a nonelectrical, solid-state source of penetrating rays in industry, to substitute for X-rays used to inspect welds in pipes and to inspect components hidden behind metal plates. There are few other examples of active isotope production reactors in the world. One is the European High Flux Reactor at

Petten in the Netherlands, and another is the OPAL reactor in Australia. OPAL began operations in April 2007.

In the early days of nuclear reactor design, some accidents were almost inevitable, as a wide investigation of reactor types, techniques, and methods was conducted in a fledgling area of expertise. The proper depth and extent of safety measures was suspected but was still unknown in a discipline that was not yet a fully functioning industry.

3 Tests Gone Awry at the National Reactor Testing Station

In the middle of World War II, in 1943, the U.S. Navy needed an isolated spot where heavy naval guns could be tested. Requirements were that the ground had to be level, with a free line-of-sight for at least 14 miles (22.5 km), so that 16-inch (40.6-cm) shells could be fired from powerful cannons, and there would be nothing on the ground that could be harmed by the detonation of a few tons of explosive. A perfect plot of land, 271 square miles (701.9 km²), was found in eastern Idaho in the Lost River Desert, and the land was withdrawn from the public domain. The nearest town, 14 miles to the northwest, was Arco, oddly named for the chief electrical engineer at Telefunken, a radio manufacturer in Germany. Construction of the Naval Proving Ground was completed on August 2, 1943, and locomotives began hauling guns, fuses, explosive charges, and projectiles to be tested out into the desert, far from inquisitive enemy eyes and ears. The ground shook when a cannon was fired, and if a seismograph had been within 500 miles (800 km), it would have registered small earthquakes.

In 1949, the need for testing naval guns had waned. Battleships had been rendered obsolete or at least second-tier by the advanced naval tactics developed in the war, but a new use was found for the remote, desolate desert facility. It was expanded and modified for testing a new technology, nuclear power. The site, increased to 890 square miles (2,300 km²), became a workplace for hundreds of nuclear technicians, engineers, and scientists, and the Atomic Energy Commission (AEC) planned to spend $500 million testing nuclear reactors of all types. The proving ground was renamed the National Reactor Testing Station (NRTS).

Over the next 30 years, 52 experimental nuclear reactors were built and tested at the NRTS, with names such as the Power Burst Facility and the Fast Spectrum Refractory Metals Reactor. Each experiment was designed to test some aspect of controlled nuclear power generation using an actual machine, running in an environment where a mishap would not jeopardize human habitation. Each test, though carefully designed for safety, had an element of risk. The tests were, after all, experiments. Most tests were military secrets, and little information concerning this work made it to the media.

THE MYSTERY OF THE SL-1 REACTOR EXPLOSION

In the early 1950s, the U.S. Army began the development of an array of over-the-horizon radar installations above the Arctic Circle in Alaska, Canada, and Greenland. The purpose of this project was to monitor the airspace over the North Pole for an invasion of bombers from the Soviet Union. It was called the Distant Early Warning, or the DEW, Line. A problem arose supplying remote radar stations with electrical power and heating, and a solution was found with small, easily installed nuclear reactors. Using nuclear power, there would be no need to keep constantly delivering diesel fuel to iced-in radar stations, and there would be no mountain of empty oil drums piling up in the wilderness.

Three types of nuclear power plant were planned, and each type had three possible power levels, so nine systems were to be designed and tested. The types were the stationary, or S, the mobile, or M, and the portable, or P. Power levels were low, medium, and high, or L, M, and H. The stationary plant was to be semipermanently installed at an arctic station but would be prefabricated and could be erected by soldiers. The mobile reactor would be a completely built unit, transportable by boat or truck, and the portable unit could be carried on the back of a jeep. For these systems, the army wanted foolproof designs. They wanted reactors that could not be damaged or destroyed by inexperienced operators and could operate in remote locations without an expert presence. For these specifications, the boiling water reactor (BWR) seemed ideal, and engineering proceeded with this in mind.

The first army reactor was the SM-1, a stationary, medium-power reactor of two megawatts electrical, built at Fort Belvoir, Virginia, 18 miles (29 km) from the White House in Washington, D.C. It was built in 1957, and it was used as a training reactor for the second army unit built, the SL-1, a stationary, low-powered reactor giving 200 kilowatts electrical and 400 kilowatts of space heating, intended to be assembled in subzero cold in Greenland. The reactor was designed by Argonne National Lab and Combustion-Engineering of Connecticut, and it was assembled in a remote location at the NRTS for shakedown testing. The entire power plant, including the reactor, the turbine, the generator, and the steam condenser, was enclosed in a cylindrical steel building, 48 feet (14.6 m) high and 38 feet (11.6 m) in diameter, resembling a short grain silo. It was not intended as a containment building as is required for any civilian nuclear plant to prevent the release of radioactive materials in a destructive accident but was simply meant to shield the machinery from the weather. Reactors are normally set in concrete, but there was no way to pour concrete in arctic conditions, because it would freeze as it was mixed and before it was allowed to harden. The reactor vessel was therefore set in a loose mixture of gravel and steel pellets. The bottom third of the building was the reactor vessel in gravel, the middle third was open space with the reactor top, the turbine, and the generator, and the top third was the "fan room," containing the air-cooled steam condenser. In the field installation, the reactor would be set upright and the gravel would then be dumped around it, in small batches. The Idaho installation followed the same procedure, as if it were in Greenland. The control room was in a separate building, connected to the reactor silo by way of a covered walk and steps spiraling up the side to a single door opening into the open floor.

The SL-1 plant went critical on August 11, 1958, and started generating electricity on October 24. It was run almost continuously, testing the systems for reliability and training several shifts of operating personnel. The experiment did not gain much attention at the NRTS, as it was not exactly breaking new ground, and there were 20 other, more interesting experiments going on in the surrounding desert. Still, the navy and the air force were both interested in the concept of small, transportable power systems and asked to participate by training Seabees and airmen to operate the reactor. The nearest installation was Rickover's S1W, five miles away, just over the horizon.

By January 1961, the SL-1 plant had run for a year and a half without serious problems, under test conditions matching the load under which it would operate in Greenland. There were some leaking pump seals. Two men from the fourth wave of operator trainees at Fort Belvoir were Dick Legg, a Seabee from Michigan, and Jack Byrnes, an army private from Utica, New York. Both were young men, hoping to work in the new nuclear power industry after leaving the military. Byrnes had joined the army at 17 and had married at 19. By the time he arrived at the NRTS to work with the SL-1, his marriage was in precarious condition and he never seemed to have sufficient operating funds. He took a moonlighting job at the Texaco filling station outside Idaho Falls.

Legg was a jokester and amused himself by tormenting Byrnes with an unexpected grab in a sensitive spot or by activating the mind-numbing scram alarm just for laughs. It was inevitable that the personalities of these two people would clash, and in May 1960 they had the inevitable drunken brawl at a friend's house. In December 1960, Byrnes was evaluated as "not ready for promotion," which meant no raise that year.

The SL-1 project was not being run with the grim efficiency of Rickover's submarine project nor with its extravagant budget. The army had only about a dozen people working on the project, Argonne Combustion-Engineering had a small crew in Idaho, and the AEC had assumed a standoff position. Interservice rivalries were interfering with efficient operation of the plant, and supervision seemed lax. An operations manual had yet to be written, the bottom of the core had started to crumble apart, and the peripheral control rods were beginning to stick. On January 2, 1961, the plant was shut down and the reactor top was taken apart to install cobalt "flux wires" in the core to measure the neutron flux shape.

At the beginning of the night shift at 4:00 P.M. on January 3, it was freezing cold, and Byrnes and Legg were assigned the job of reassembling the top of the reactor. Joining them on this night was Richard McKinley, a trainee from the air force who was supposed to watch, learn, and assist as Byrnes and Legg worked on the reassembly task. According to the work log, progress was slow and behind schedule. At 7:00 P.M. Byrnes's wife called to tell him that she was filing for divorce, he should find other quarters, and she would be withdrawing half of his paycheck, which had yet to arrive in the mail.

At 9:01 P.M., an automated radio message, one long tone and two shorts, was heard in the NRTS fire stations and the security center. The coded signal meant that SL-1 was on fire, and the Central Station was closest to the site. The men at Central read the card containing special instructions for the SL-1 as they suited up, climbed on the fire truck, and rolled out into Highway 20. Nine minutes later, they arrived at SL-1, and a detail ran to unlock the gate. It was bitterly cold, and a small wisp of steam above the silo was visible in the floodlights.

The scene was peaceful to the point where it seemed strange. There was no smoking wreckage, no glow of flames, and no one running out, screaming, to meet the truck. It looked like a false alarm, which was not too unusual, but the place also seemed deserted. There was supposed to be a crew at the reactor 24 hours a day. The firemen knew to approach any reactor building at NRTS only with portable radiation counters in hand. They went first to the control room. No one was present, but radiation alarms were going off all over the building. They next took the covered walkway to the silo and started up the stairs. The radiation counters began to register extremely high radiation, with the instruments going off the highest scale at 500 rads/hour. They turned around and came down the stairs.

By 9:17 P.M., a health physicist (HP) from a neighboring experiment had arrived. He suited up in a face-covering respirator with an air tank and started up the stairs on the silo. Seeing his radiation counter go off scale, he retreated. The entire site had been searched by this time, and although there were supposed to be three men on duty, they were not to be found. They had to be in the reactor building. The assistant fire chief and two HPs then found a more robust portable radiation counter, dressed in full radiation-protection clothing, and climbed the full stairway to the silo entrance. They confirmed that the radiation dose rate was 500 rads/hour. At that level of exposure, they had to limit their time at the silo door to a few seconds. The chief opened the door and glanced in. He saw destruction. The reactor's water, which should have been confined to the cooling loop, was on the floor, mixed with loose gravel and steel pellets. The reactor top was a mass of twisted metal. His face mask fogged in the below-freezing temperature, and he could not see the three men.

By 10:30 P.M., Ed Vallario, the HP in charge of SL-1, and Paul Duckworth, the head engineer from Argonne Combustion-Engineering, had arrived. By this time the situation looked very bad. Vallario and Duckworth suited up and took the stairway, intending to find the crew. As they entered the silo, their radiation counter registered 1,500 rads/hour. At that dose rate, 20 minutes of exposure could be deadly, so they had to hurry. Walking out onto the floor of the open room, they saw Byrnes lying on the floor, bleeding from massive injuries, moving slightly, and moaning. McKinley was

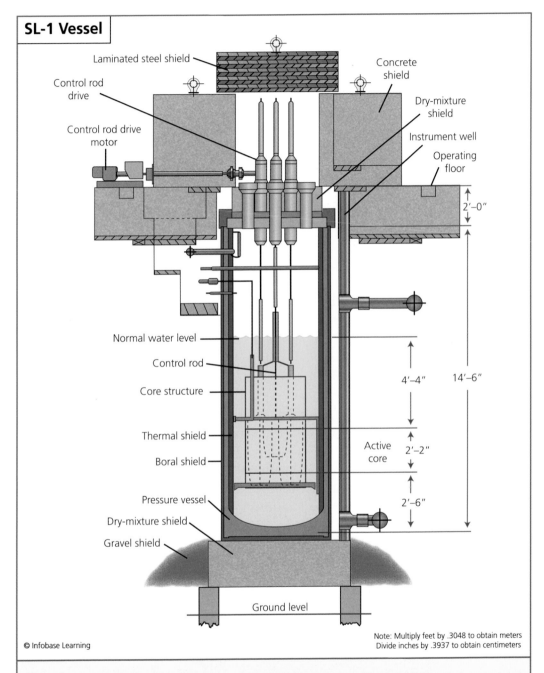

SL-1 Vessel

Laminated steel shield

Control rod drive

Control rod drive motor

Concrete shield

Dry-mixture shield

Instrument well

Operating floor

2'–0"

Normal water level

Control rod

Core structure

Thermal shield

Boral shield

Pressure vessel

Dry-mixture shield

Gravel shield

4'–4"

14'–6"

Active core

2'–2"

2'–6"

Ground level

Note: Multiply feet by .3048 to obtain meters
Divide inches by .3937 to obtain centimeters

© Infobase Learning

The internal structure of the SL-1 reactor core, showing the massive centrally located control rod that could turn the reactor on and off with one motion.

nearby. He had apparently been blown against a large, concrete block by an explosion at the center of the floor, and he had died instantly. Legg was nowhere to be found. Vallario and Duckworth retreated down the stairs as quickly as they could move in protective clothing.

The personnel at the NRTS were trained, drilled, and equipped for emergencies involving the frontier of nuclear power, but there was no training or planning for this unanticipated occurrence. A mission to retrieve Byrnes was improvised quickly. Five men would storm the building, two with a stretcher and three holding stopwatches and radiation counters. No one would be allowed in the building for more than 60 seconds. In that time, they would have to cross the floor, pick up Byrnes, toss him onto the stretcher, and make it back to the door.

The five climbed the stairs, somewhat impeded in their radiation suits, and went through the door. The plan seemed simple, but a difficulty was walking on the gravel. The coolant covering everything had begun to turn to slippery ice. The stretcher-bearers managed to pull Byrnes down the stairs, but then they realized that he was part of the reason why the radiation dose in the building was so high. Byrnes alone was a source of 1,000 rads/hour of gamma radiation, and it was dangerous to be in his presence.

Byrnes stopped breathing at 11:00 P.M. There was nothing medical that could be done for him, so to cut short the radiation exposure the ambulance driver turned out into the empty desert, drove hard for a minute, then dove out of the vehicle and ran back to the road. He and the NRTS night nurse had received significant radiation doses.

At this point, the SL-1 problem was turned over to the AEC, and it was a puzzle and a fatal accident of unprecedented severity. Their immediate tasks were to retrieve McKinley's body and to locate Legg and to do so required mission planning, full-scale mockups of the SL-1, training and rehearsals, and hundreds of personnel. Legg was eventually found, not on the floor with the other two crewmen, but pinned to the ceiling by part of the central control rod piercing his body. He was removed by a specially built, remote-control crane sticking through the door, nine days after the incident. All three men were heavily contaminated with vaporized fuel, laced with a year-and-a-half accumulation of radioactive fission products. They had to be buried in lead coffins to prevent radiation contamination of burial sites. Selected body parts had to be treated as high-level radioactive waste and were buried in sealed drums.

The second stage of the problem was to determine what had happened on the night of January 3, 1961. The reactor had exploded from the coolant having flashed to steam while three workers were reassembling the

The SL-1 core was heavily contaminated with radiation, and it had to be removed by remote control and buried in the desert. *(Idaho National Laboratory)*

control mechanism. The metal silo, while it was never intended to seal the reactor against radioactivity leakage, had done a good job of containing the explosion. None of the wrecked fuel had escaped, but a plume of gasified fission products, including iodine-131, had blown away up to 100 miles (161 km) to the southwest. What was not known was why this had happened. The BWR concept was well known as an inherently safe reactor, and it should have been impossible to make it explode under the worst circumstances. Although it had not melted down or caused serious contamination of the surrounding desert, in less than a second it had killed three men and destroyed itself.

A thorough, two-year investigation was mounted. No expense was spared. The accident was staged in full-scale recreations. Using computer-code simulations, individual pieces of the reactor were microscopically examined,

Health physicists check Highway 20 for radiation spread by the SL-1 reactor accident (*Idaho National Laboratory*)

and an animated, color film was made, showing the top of the reactor come apart in slow motion. From these analyses, it is known that the workers were in the process of reattaching the central control rod to its drive motor. The reactor was, of course, in shutdown mode, with all control rods parked in the fully inserted position, preventing fission from occurring in the reactor core. The center rod was 14 feet (4.3 m) long and weighed 100 pounds (45.4 kg). The uranium-fueled core was 12 feet (3.7 m) down, under their feet and covered with cooling water. The center rod had been lifted slightly and had a C-clamp screwed down on it.

Legg was squatting down over the bell housing, where the control rod protruded from the metal reactor top, which was bolted down on the top rim of the reactor vessel. Byrnes was standing over him, with his legs bent, holding the end of the control rod with both hands, and McKinley was standing to the side, off the reactor top, watching. Byrnes was supposed to lift the rod one inch (2.54 cm) to take the load off the C-clamp, at which time Legg would unscrew it and set it aside. For some unknown reason, Byrnes jerked the rod out by 23 inches (58.2 cm). It was only necessary to pull the rod 16 inches (40.6 cm) to make the reactor go critical. Moving it rapidly to 23 inches (58.2 cm) caused the reactor to go critical on prompt neutrons alone and suddenly generate 19 billion watts of pure heat. The fuel vaporized, and the water surrounding the two-foot- (61-cm) high uranium core flashed instantly into steam. The steam drove the water sitting on top of the core up through seven feet (2.1 m) of empty space in the reactor vessel until it hit the top, like a sledgehammer. The impact was great enough to shear away the water pipes welded to the reactor vessel and send it flying out of its gravel bed. It hit the ceiling of the room and then fell back into the gravel. The incident was mapped in millisecond increments around T, at 9:01:00 P.M.

At T minus 0.500 seconds, Byrnes begins to pull the rod.

By T minus 0.120 seconds, Byrnes has overshot his target withdrawal and pulled the rod to 16.7 inches (42.2 cm). The reactor goes critical.

At T minus 0.0, the rod has been withdrawn to 23 inches (58.2 cm) and the reactor is promptly supercritical. Power peaks at 19 billion watts in a reactor that was designed to run at 600 kilowatts. The fuel temperature goes to 3,740°Fahrenheit (2,060°C). The melting point of the uranium-aluminum alloy fuel is 1,292°Fahrenheit (700°C). The fuel, along with the embedded fission products from a year and a half of power production, turns to vapor.

At T plus 0.0005 seconds, the BWR self-scrams and goes subcritical, as the water moderator has left the core dry. The column of water above the core starts to move upward.

At T plus 0.034 seconds, the water hits the lid of the reactor vessel and causes 10,000 pounds (44,482 Nt) of force on it. Parts begin to fly off the top of the reactor, traveling at 85 feet per second (26 m/s). The reactor tank, 13 feet (4 m) long, starts to move.

At T plus 0.160 seconds, the first fragment of the reactor hits the ceiling, and two-thirds of the water has left the vessel.

At T plus 0.800 seconds, the top of the reactor hits the ceiling.

By T plus 4.000 seconds, the reactor has fallen back into its pre-explosion position, and the event is over.

Even with all this meticulous analysis of the accident, there is still no satisfactory explanation for why it occurred. Everyone who had ever taken the reactor operator course at Ft. Belvoir was made thoroughly aware of the danger of rapidly ejecting the controls. There had even been idle speculation among students as to what they would do to deny a reactor to the Soviets if they were to invade and take over a DEW Line radar station. They had decided that the best thing to do would be to jerk out the control rods and blow the reactor sky-high.

Did Byrnes accidentally jerk out the central rod? The operations log showed that rods had been sticking in the bottom of the core and would not move smoothly. Perhaps he thought he would exercise the rod while they had it disconnected from the motor by working it up and down. If it were stuck, it could have suddenly released with Byrnes exerting greater than normal upward force. The major problem with this scenario was that the peripheral rods had been sticking, but never the central rod.

All possibilities were investigated, no matter how unlikely. Had the prankish Legg grabbed Byrnes in a sensitive spot as he was pulling up the heavy control rod, and did this make him jump and jerk the rod out of the reactor core? This possibility was given more than serious consideration, as multiple volunteers were asked to stand over a control rod bell housing and pull a rod of appropriate weight up one inch (2.54 cm). Completely by surprise, they were then grabbed by a man crouched below. In no case did a surprised individual pull up the rod. Most dropped it.

It may be that Byrnes, distraught over recent deteriorations in his life situation, committed suicide atop the reactor, and in doing so he murdered his two fellow workers. The reason for the SL-1 incident remains a mystery. It is still the worst disaster in the history of the U.S. nuclear power program. The design of the BWR, and all other reactor designs, was modified due to the destruction of the SL-1, so that this type of accident involving nuclear power can never happen again. The SL-1 was buried in the desert, 1,600 feet (488 m) from the site of the experiment at coordinates Lat: 43.5174820, Lon: –112.8228310.

Army reactors ML-1 and PL-1 were successfully tested without incident, and PL-1 went on to power an automated weather station on Axel Heiberg Island, north of the Canadian mainland. SL-1 reactors were put into service near both the North and the South Poles. The army reactor program was cancelled in 1965, as funding was diverted into military action in Vietnam.

IMMEDIATE CHANGES IN REACTOR DESIGN PRINCIPLES FROM SL-1

The fatal flaw in the SL-1 reactor design was the fact that it could go critical with the motion of a single centrally located control. Beyond that problem, the worth of the central rod was such that supercritical fission could be accomplished without the need for delayed neutrons, meaning that the reactor could go promptly critical. The near instantaneous increase in power brought on in prompt criticality is impossible to control with mechanical systems. Although the reactor will always go subcritical under such a condition, destruction occurs in the few milliseconds before subcriticality is reached.

Few experimental reactor designs were designed this way, with most having the reactivity worth divided among several controls. The SL-1, however, was an economical design, meant to be as simple as possible, so the number of controls was kept to a bare minimum. There were four other control rods, but they were located on the periphery of the core and were mainly used to flatten the flux shape so as to even out the fuel burn-up. The SL-1 was also designed to go three years between refuelings, so the reactivity of the uranium core was made as high as possible. After running three years, all five controls rods would have to have been pulled to make the reactor critical.

After analysis of the SL-1 incident, no nuclear reactor was designed with one control capable of initiating criticality. The procedures to connect control rods to the drive mechanism were changed, making it impossible to manually pull a rod out of the reactor core. Operational control motion was changed. It is no longer possible to pull out a control with a single motion from the control room. Controls are moved only in small increments. One 90-degree twist of a handle on the control panel bumps a rod by only an inch (2.54 cm). The design goals of simplicity and economy are now approached with the greatest care, never compromising safety of operation or maintenance. Since the SL-1 incident, no power reactor has been built in the United States without a containment building surrounding the reactor system, capable of preventing the escape of radioactive contamination in the event of an explosion.

Problems encountered and solutions that were developed after the SL-1 incident changed the way the AEC and later the Nuclear Regulatory Commission (NRC) planned for nuclear accidents. It provided a guide and a benchmark for the study of the spread of radioactive contamination after a nuclear power accident. The impor-

(continues)

(continued)

tance of radiation dose measurements of all personnel involved with the accident became clear, as well as environmental monitoring and the use of photography, both aerial and close-up, of the accident scene. The maximum range of hand-held radiation detection equipment was increased to prevent unknown dose rates in high-radiation environments. Respirators were modified so that they could not freeze and fog up in cold weather. Smoke pots were specified for future use, so that wind currents near an accident site could be monitored continuously. Calculation methods for radiation dispersal have been improved, and mechanisms for teaming with the U.S. Weather Bureau are in place for tracking airborne radiation.

A safety poster was printed to be hung in every nuclear engineering office in the United States. It shows a photograph looking down into the twisted wreckage of the SL-1 reactor vessel and on it is written LEST WE FORGET. SL-1. 1-3-61.

The only fatal nuclear reactor accident in the United States occurred here, atop the SL-1 power reactor in Idaho. This photo was later turned into a safety poster for nuclear engineers. *(Idaho National Laboratory)*

THE EXPERIMENTAL BREEDER REACTOR NUMBER 1 MELTDOWN

The concept of a breeder reactor, or a reactor that can produce more fuel than it uses during normal operations, was very attractive in the early days of nuclear power experimentation. After World War II, there was a push to move nuclear technology beyond weapons manufacture and into power production, making electricity with a new method that did not require burning coal or petroleum. The one glaring problem with this plan was a lack of uranium fuel in the United States. The main source of uranium was the Belgian Congo, Africa, and there was no guarantee of continuous service from these foreign mines.

In theory, it was possible to build a reactor that would run on either uranium-235 or plutonium-239. Both isotopes are *fissile*. Uranium-235 is extracted from natural uranium at great expense. Plutonium is made artificially, by bombarding uranium-238 with neutrons. There was no shortage of uranium-238, and great stockpiles of it were left over from the uranium enrichment program of the late war. Over 99 percent of natural uranium is composed of the non-fissile uranium-238. The process of plutonium production had been proven during the war, as large graphite-moderated reactors were used to produce excess neutrons and convert uranium into plutonium for use in atomic bombs. It could be proven on paper that a reactor could be built to produce heat for steam generation and power production and excess neutrons could be captured by a blanket made of uranium-238 surrounding the core. Under the right conditions, such a reactor will convert non-fissile uranium-238 to fissile plutonium-239 at a rate greater than plutonium-239 is being consumed as fuel. In this configuration, the reactor is not simply converting plutonium, it is breeding it. If the power network in the United States were equipped with breeder reactors, then there would be a nearly endless supply of fuel for a process that generates no carbon dioxide.

Walter Zinn of the Argonne National Laboratory in Illinois proposed a breeder reactor experiment for the NRTS in 1949, and it was built in a blank spot on the map of the Idaho desert, out of sight of any other reactor experiment. It was named Experimental Breeder Reactor Number One *(EBR-1)*, and its purpose was to validate the physics theory predicting that a breeder reactor was possible.

There was a wrinkle in the breeder reactor theory. To convert plutonium at a rate such that more was produced than was burned, the process has to be optimized. Plutonium conversion is most likely to occur at high neutron speed and not the slow, *thermal speed* at which all normal reactors operate. Fission probability is optimized at low neutron speed, and plutonium conversion is optimized at high neutron speed. By increasing the amount of fissile fuel beyond what is needed under the usual conditions of slow neutrons, a reactor can be made that depends only on fast neutrons for fission. This configuration is called a fast reactor, for the speed of the fission-producing neutrons. A fast reactor must have a coolant. The coolant in a normal reactor is something lightweight, such as hydrogen in water, or even deuterium, and the coolant acts doubly as a moderator to slow down the neutrons to fission speed. In a fast reactor, a coolant is needed that will carry heat away from the fuel without slowing down the neutrons. The coolant must therefore be heavy, or higher up on the periodic chart of the elements. Molten metal is the best coolant for a fast reactor, and most use sodium or a mixture of sodium and potassium metal. The engineering problems of using metal that is solid at room temperature for a coolant that must cycle unimpeded through a reactor core and a heat-exchanger are challenging. The EBR-I was built not to prove that plutonium could be bred, but to prove that a reactor could be cooled by liquefied metal.

Unlike a water-cooled, or thermal, reactor, the fast reactor would not necessarily go into immediate shutdown mode if the coolant were lost. A fast reactor lacks the wealth of delayed neutrons in a thermal reactor, and this makes it touchy at the controls. The reactivity can increase, or decrease, with coolant loss or with temperature change, and this will either cause the power level to increase or decrease, depending on many factors that must be juggled. It was this uncertainty of effects that the EBR-1 was built to examine and illuminate. It was built using highly enriched uranium metal fuel, formed into tiny pellets and canned into long rods made of stainless steel, less than one-half inch (1.3 cm) in diameter. The total core mass was 115 pounds (52 kg), bathed in a stream of sodium-potassium, or NaK. Construction was in 1951, and EBR-I first went critical, at very low power, on August 24, 1951. Tests were successful, and the heated NaK was used to produce steam in a water loop. At 1:50 P.M. on December 20, 1951, the steam was used to spin a small turbine connected to an electric generator, and the power produced was used to light four 200-watt lightbulbs strung across the turbine deck. This was recorded as the first electrical power ever produced by nuclear means, and it was from a reactor breeding plutonium-239. The breeding action was confirmed by chemical analysis of the uranium-238 blanket in 1953.

As was the case for every nuclear reactor ever built, EBR-1 had a scram system designed to stop the self-sustaining fission action and shut it down, either automatically on detection of runaway conditions or by manual actuation of a large red button on the control panel. Thermal reactors usually scrammed by dropping the entire control complement into the core, by both spring action and gravity. The EBR-I scram dropped away the breeding blanket, which was a heavy cylindrical shell of uranium-238, by gravity. Most of the neutrons streaming out of the core and into the

The first electrical power ever generated by a nuclear reactor was here at the EBR-I turbine deck. *(Idaho National Laboratory)*

breeding blanket were captured by the uranium-238, but a percentage of them would bounce off and back into the core, where they would contribute to the fission process. The core could be shut down by simply denying it these reflected neutrons.

On November 29, 1955, EBR-I was under test to find whether the power level would rise or fall if the coolant flow were stopped. This was one of a long series of tests in the unknown region of fast reactor response under temperature or coolant perturbations. The reactor was made critical and taken to a very low power level of 11 watts. All systems were normal, and the power was increased to 50 watts. The controls were then set to increase the power with a period of 60 seconds, meaning that the power level would increase by a factor of 2.72 every passing minute.

The goal was a power level of 500 watts, at which point the controls would be adjusted, bringing the power down to stable operation with a period of zero.

Three seconds later, the power was 1 million watts. The reactor was completely out of control. The scientist in charge yelled for the operator to scram the reactor. The operator, by mistake, hit the button that would drive in the control rods, instead of the scram button. The in-traveling controls were shutting down the reactor, but not quickly enough. After a delay of two seconds, during which power indicators ran off scale, the chief scientist lunged for the scram button and dropped the blanket. The power level had peaked at about 10 million watts, which was much higher power than the reactor systems were designed to handle. Radiation was detected in the cooling system, indicating that the fuel cladding (metal coating covering the core) had failed.

None of the personnel at the EBR-I that day were dosed with radiation, but the core was almost completely melted. The unusually high core temperature from the lack of coolant flow had caused the fuel rods to bow in slightly, and this small change in the core configuration had improved the reactivity and caused a runaway fission reaction. Computer simulations of the accident indicated that if the scram had not been activated the power increase would have been 2.5 times higher, but that the reactor would not have exploded. EBR-I was dismantled and rebuilt, continuing research into the dynamics of liquid-metal-cooled fast breeder reactors until 1964. It was the first reactor to run with a pure plutonium-239 fuel load. On August 25, 1966, EBR-I was granted U.S. National Historic Landmark status, and it is now a tourist attraction in Idaho.

EBR-I was replaced by EBR-II in 1965, running at 62.5 megawatts. EBR-II ran successfully for 30 years, and much was learned from it concerning a variety of metallic and ceramic fuel configurations.

4 The Sellafield Facility in the United Kingdom

Immediately after World War II, Great Britain found itself abandoned in a new world order of nuclear weapons and veiled threats of aggressive movement in Eastern Europe from the Soviet Union. British scientists had participated openly in the crash development of nuclear weaponry, the Manhattan Project, working hand-in-hand with the United States and its recently inherited group of European expatriate nuclear scientists. However, everything was not shared with the British during the bomb project, and that fact was noted by then prime minister Winston Churchill. While their help was gratefully accepted at the Los Alamos, New Mexico, bomb laboratory, Canadians and Brits were not allowed near the plutonium production facilities at the Hanford Works in Washington State or the uranium enrichment plants at *Oak Ridge* in Tennessee.

Churchill met with Franklin Roosevelt, president of the United States, in Quebec City, Canada, on August 19, 1943, to work out an agreement in which the United States, Canada, and the Britain would promise to share all technical information concerning nuclear fission after the war had ended. The Quebec Agreement was finally signed on September 18, 1944, at Roosevelt's home in Hyde Park, New York.

To the dismay of Churchill, the United Kingdom, and much of the Western world, President Roosevelt died of natural causes on April 12, 1945, and the Quebec Agreement died with him. The document seemed to have been misplaced, and the new president, former vice president Harry

S. Truman, was not of a mind to share any nuclear secrets with anyone, regardless of how helpful they may have been during the war. In August 1946, the U.S. Congress passed the Atomic Energy Act, making it forbidden to pass information concerning nuclear fission or its related technology to any foreign country, including Great Britain. President Truman quickly signed it into law.

The Soviet Union, with an extraordinarily well-developed espionage network within the nuclear weapons research and production facilities in the United States, was rapidly developing its own atomic bomb. The United Kingdom feared another war in which Europe would be overrun, this time not by Nazi Germany but by Communist Russia, using the newly invented nuclear technology. In near panic mode, Great Britain had to establish its own atomic bomb production, beginning with a plutonium production facility. Plutonium was the least expensive fissile ingredient for an atomic bomb, and it could be made by exposing otherwise inert and plentiful uranium-238 to neutrons produced in a nuclear reactor. High power at the megawatts level was necessary to accomplish this conversion.

Great Britain had many competent nuclear scientists and a technical, industrial infrastructure but lacked a resource plentiful in the United States: desolate, uninhabited deserts. The United States had wide-open spaces for bomb testing and unused rivers for production reactor cooling. Dangerous nuclear experiments could be conducted far from human habitation. Where the United States had deserts, Great Britain had abandoned air bases. In 1945, John Cockcroft, who had worked on the Manhattan Project in New Mexico, was asked to establish a research laboratory for studies of nuclear fission for military and commercial purposes. He formed the Atomic Energy Research Establishment (AERE) on January 1, 1946, at the Royal Air Force Harwell airfield, 16 miles south of Oxford, England. Existing buildings were put to immediate use, housing nuclear experiments, scientists, and engineers.

Small research reactors were built at Harwell, but the powerful production-scale reactors needed a more isolated location. The site chosen was the Royal Ordnance Factory (ROF *Sellafield*) on the Cumbrian coast, overlooking the Irish Sea. During the war, a great deal of TNT high explosive was manufactured there, and the near constant wind blew the industrial smoke away from land. It was named *Windscale,* for a bluff overlooking the seaward side.

This chapter covers accidents at Windscale, most of which were minor and not unusual for a large nuclear weapons production facility. One incident, however, a fire in the Unit 1 plutonium production pile, was very serious. It was, in fact, probably the third worst nuclear power disaster in history, eclipsed only by the Chernobyl calamity in the Ukraine in 1986 and the Fukushima I earthquake incident in Japan in 2011. Afterward, the facility found that its name was forever connected with this accident, so it was changed to Sellafield, as it is known today. These nuclear accidents at Windscale are detailed, and a sidebar discusses the local economic and health effects of living in the vicinity of a major nuclear incident.

THE WINDSCALE FIRE

The first nuclear reactor built in Europe was the Graphite Low Energy Experimental Pile (*GLEEP*), erected at Harwell in an airplane hangar. The design was primarily based on what Cockcroft had managed to learn about the X-10 graphite reactor at Oak Ridge, Tennessee. GLEEP was air-cooled and consisted of 11,500 metallic, natural uranium fuel slugs, each about the size of a roll of quarters, clad with aluminum. These fuel elements were assembled into aluminum rods, which were then loaded into 676 holes bored horizontally into the two-story-tall cube of graphite bricks comprising the reactor. The reactor was first brought to power, operating at only three kilowatts, on August 15, 1947. This low-power experiment was necessary to learn how to build a high-power production reactor using the same technology, but the GLEEP remained active for an exceptionally long 43 years, becoming an international standard for materials testing and neutron measuring instrument calibration.

With extreme urgency, work began in September 1947 on a pair of plutonium production reactors at Windscale, even before there was sufficient knowledge to design them. Based on the successful operation of GLEEP, a more powerful reactor, the British Experimental Pile 0 (BEPO), was built in 1948 to find further design parameters for the air-cooled reactors. Cooling fins attached to the aluminum fuel cladding proved sufficient to keep the uranium from melting with air blown through the pile at atmospheric pressure, and the air-cooled power reactor design was deemed feasible. Construction at Windscale proceeded in parallel to the BEPO work at Harwell, employing nearly 5,000 construction workers and 300 architects, engineers, and surveyors. Although the reactor was designed to be as simple and economical as possible, there were still unknowns and engineering difficulties. A fuel-loading machine, intended to slide fuel cartridges into the reactor face by remote control, proved impossible to build, and the task was simplified by having men stand in front of the pile and push fuel in, using sticks.

Although U.S. scientists were forbidden to share information with anyone, there was sympathy for the British effort, and in 1948 critical data concerning graphite was given to Cockcroft's team. In the large water-cooled graphite reactors at Hanford, effects were observed confirming predictions from Dr. Eugene Wigner of the Los Alamos Laboratory. Wigner had postulated a disturbance in the crystalline lattice of graphite, caused by high-energy neutrons colliding with carbon atoms, literally knocking them askance. This sub-molecular collision causes two effects. The "Wigner growth" is a drifting of the dimensions of a graphite block, due to the crystal disruptions. Allowance must be made for the size of the reactor to grow and for critical passages for control rods to shrink or warp. The "Wigner energy" is the potential energy stored in the graphite by having its carbon atoms displaced. If allowed to accumulate, this energy can let go all at once, causing the graphite to catch fire. It must be annealed out of the graphite on a periodic basis, by raising the temperature to a specified level and then lowering the temperature slowly to allow the Wigner energy to dissipate. These effects were not obvious in any low-power experiment and only came about when the Hanford piles were operating at full power. Cockcroft was also advised that a graphite fire could be extremely difficult to extinguish. Pouring water on it only makes the fire hotter, as the oxygen is scavenged out of the water, leaving hydrogen gas. The hydrogen gas is then a secondary hazard, combining with atmospheric oxygen and causing an explosion. Special care must be exercised in using a graphite pile at high power.

The Windscale piles were massive structures, each built using 1,968 tons (2,000 mt) of graphite in a reclining, octagonal semicylinder, 24 feet (7.3 m) long by 50 feet (15.2 m) in diameter. Each of the 70,000 fuel cartridges in one reactor was one foot (30 cm) long by one inch (2.5 cm) in diameter, clad in finned aluminum cans. The bio-shield, built to protect any personnel and the surrounding countryside from radiation, was concrete, seven feet (2.13 m) thick. Air was sucked in on both sides of the building by eight electrically powered blowers, blown through channels cut in the graphite core to cool the reactor core, and then sent up a stack 410 feet (125 m) high. From there the air, which was probably made radioactive by its trip through the fissioning core of the reactor, was blown out over the Irish Sea by prevailing wind.

As the buildings were being completed in late 1948, Cockcroft returned from a fact-finding mission to the United States with sobering news. If one of those 70,000 fuel cartridges were to break open, then radioactive fission products would go up the stack. If the wind were not blowing to the west, then highly radioactive dust would come down over the farm country surrounding the Windscale facility. The air would have to be filtered before it went up the stack. This late-breaking design change did not sit well with the men who had just spent a year laying concrete. There was no spare chamber in the building where a filter big enough to accommodate the airstream could fit. Cockcroft was insistent, and he was the head of the AERE, so the engineers backfitted fiberglass filters to the tops of the two Windscale air stacks. The labor required was tremendous, with 200 tons (181 mt) of structural steel and concrete hauled to the tops of the stacks. The newly assembled filters were named "Cockcroft's follies."

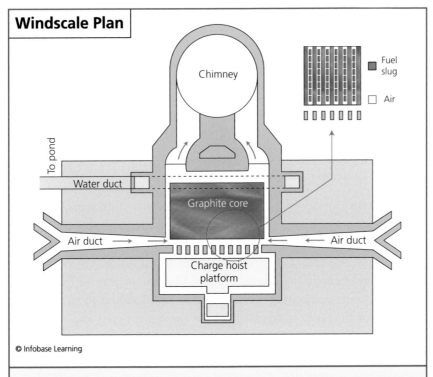

Windscale Plan

Chimney

Fuel slug

Air

To pond

Water duct

Graphite core

Air duct →

← Air duct

Charge hoist platform

© Infobase Learning

A plan view of the Windscale plutonium production reactor, showing the path of air from outside the building, through the graphite core, and up the chimney

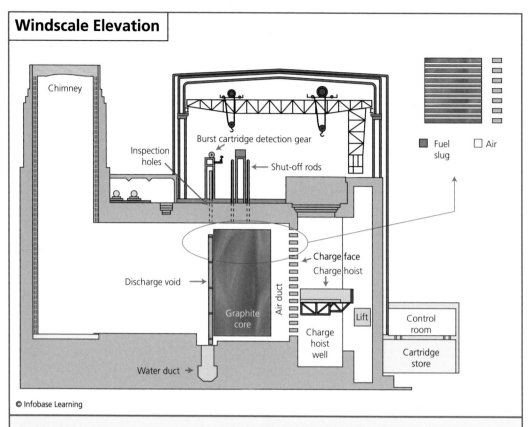

Windscale Elevation

An elevation view of the Windscale plant, showing the charge hoist on which men stood using poles to push burning fuel through the reactor. If free to move, the fuel cartridges would come out the back side of the reactor and fall into the water duct.

Even with the delays caused by helpful information leaks from the United States, the Unit 1 pile was about ready for start-up in the early summer of 1950, but Cockcroft had found one more stumbling block. New calculations indicated that the reactor could not go critical with its present fuel loading. The *critical mass* of the reactor was off by as much as 250 percent. Something would have to be modified to increase the reactivity of the core, preferably without tearing it down and rebuilding it larger. A reactor using natural uranium must be optimized in every way to make fission possible, and the purity of materials is particularly important. Anything that will parasitically absorb neutrons without producing fission must be eliminated. The engineers finally found one structural element that could be modified: If the fins on each of the 70,000 fuel cartridges could be shortened by one-sixteenth of an inch (0.2 cm), then the reactor could produce power. In three weeks of hard labor, workers at the Unit 1 pile trimmed 1 million fins.

Windscale Unit 1 went critical in October 1950, followed by Unit 2 in June 1951. On March 28, 1952, the first sample of British plutonium-239 was refined from fuel in Unit 1. With some help from plutonium production in Canada, the United Kingdom was able to construct and test its own atomic bomb in shallow water just off the coast of western Australia on October 3, 1952. The Operation Hurricane detonation, set off in the hold of an outdated British navy frigate, the HMS *Plym,* yielded the energy of 25 kilotons of TNT (100 trillion j) and dug a shallow crater in the seafloor.

Plutonium production with Windscale Units 1 and 2 continued at full capacity. Small problems developed, including a few burst fuel cartridges, inexplicable core-temperature changes, and uneven annealing of the graphite. A robotic device, the burst cartridge detector, was built to scan the face of the reactor and stick a small nozzle into fuel passages, sniffing for unusually high radiation, but unfortunately it tended to punch holes in the fuel cartridges. By October 1957, Unit 1 was due for its graphite to be annealed for the ninth time.

Annealing had become a routine exercise. The reactor core, which normally ran at a temperature between 122° and 176°F (50° and 80°C) was run up to 482°F (250°C) by turning off the air blowers. The core temperature was monitored at 66 key locations, and it was possible to detect the controlled Wigner energy release as the pile was then gently cooled by turning up the blowers. On Monday, October 7, 1957, the annealing began in routine manner at 11:45 A.M.

In addition to the operating staff and hundreds of workers at Windscale, there were five key men who would find their lives changed by this ninth annealing. H. G. Davey was the works general manager, in ultimate charge of Windscale. Ian Robertson was the pile physicist, and he was feeling sick and woozy. England was suffering from a flu epidemic at the time, and the virus had sickened many people. He was developing a fever. Tom Tuohy (1917–2008) was the deputy works general manager, a chemist educated at Reading University who had worked at ROF Sellafield during the war. Tuohy's entire family was sick with the flu, and he was trying to stay home to care for his wife and children. Tom Hughes was the acting works manager. Ron Gausden (1921–97), an engineer from Borough Polytechnic, was the shift supervisor of operations. Huw Howells was the health physics manager, in charge of monitoring radiation levels at the Windscale operation.

After switching off the main blowers, the operators began slowly running out the control rods at the lower section of the core to begin nuclear heating. At 1:00 P.M., the rod motion was paused to note core temperatures, and motion was resumed at 2:15 P.M. with all the blowers turned off. Abnormal temperature readings were indicated. The rods were stopped, and a bad set of electrical connections was found. Control rod withdrawal was resumed as soon as the temperature readings were explained, and at 7:15 P.M. the pile achieved critical, self-sustaining fission. The Wigner effect seemed localized at the bottom of the core, so the rods were adjusted to concentrate the heat in the lower area.

At 1:00 A.M. on Tuesday, October 8, two of the 66 thermocouples indicated 482°F (250°C), indicating the start of annealing. The operators shut down the fission to allow the annealing to naturally spread throughout the reactor. Ian Robertson, the pile physicist on duty, was not feeling well, and he went home at 2:00 A.M. The high temperature was supposed to spread to the entire core, but instead the temperature seemed to be dropping by 9:00 A.M. Robertson was back at Windscale by then, trying to figure out what was wrong.

The reactor controls did not feel right. The core seemed touchy, with small increments of control rod position causing larger than normal power changes. One thermocouple showed a sudden temperature increase to 716°F (380°C), and it was hard to bring it down using the local control rod. Robertson had to go home, and an assistant took over. At 11:00 A.M., the reactor was once more brought to criticality to try and restart the annealing. At 7:25 P.M., the operators shut it down once more, expecting the Wigner effect to heat the graphite.

The process seemed to be working on Wednesday, October 9, as temperature was rising with the reactor shut down. In the afternoon, the temperature suddenly took off, rising to 779°F (415°C), and that was not normal. There was still no need to panic, and the operators closed the inspection ports on top of the reactor room and shut the hatch at the base of the air stack. After letting the pile settle down, at 10:15 P.M. they opened the blower dampers to let some air flow, and the core temperature started to drop.

Right after midnight, on Thursday, October 10, the temperature started rising again, up to 752°F (400°C). Ron Gausden was in charge of operations on this shift, and he saw that something was wrong but was not sure what it was. The operators once again opened the blower dampers, but this time with no effect. By 2:15 A.M., one thermocouple showed the temperature still rising. Going strictly by the reactor operator's manual, they reopened the blower dampers. The temperature started to go down, so they closed the dampers, but then the temperature climbed again. This was very abnormal behavior. At 5:10 A.M., the operators let more air into the pile, and the radiation-measuring instrument at the top of the air stack started indicating unusual activity. Shortly after that, radioactivity was measured on the roof of the Windscale meteorological station.

The operations crew followed the operating instructions to the letter and opened air dampers at 12:00 A.M. and 1:30 P.M. for 15 minutes of cooling. This time, the air stack radiation monitors went to alarm mode. They had apparently burst a fuel cartridge and were sending fission products up the stack, so Gausden directed the operators to turn on the blowers, in shutdown mode. At 2:30 P.M., Gausden's concern had turned to urgent alarm, and he ordered that every blower be turned on to highest speed. "Blow the pile cold," he said. He hoped to use the burst cartridge detector to find the leaking aluminum can, but it was not designed to operate at annealing temperature. Gausden rang up Tom Hughes, the acting works manager, to tell him that they had a problem in Unit 1. Shortly after the call, he was contacted by Huw Howells, the health physicist, who asked him if anything unusual was happening at Unit 1. The situation began to sink in to Gausden, and he called H. G. Davey, the general manager, to report his concerns.

A crew of three workers was sent to the reactor face with a radiation counter to see if they could locate the leaking cartridge. A movable floor section in front of the reactor, called the "charge hoist," gave workers access to any fuel channel by allowing them to walk left and right across the face and move up and down with a powered lift. They opened the cover on a likely fuel channel and were shocked by what they saw. The fuel was lit up, bright red. Metallic uranium burns readily in air, which is one reason for the aluminum can surrounding a fuel element. One had broken open and caught fire. They removed another plug and saw the same thing—then more plugs. All channels were glowing red. It was worse than a fuel cartridge being on fire. The reactor was burning, and it had been on fire since Tuesday. On the far side of the reactor was a water trough, made to catch hot, expended fuel as it was pushed through its channel from the front for transport to the chemical plant for plutonium extraction. A crew member took a long, steel pole and tried to slide out one of the burning fuel cartridges. It would not budge. It was jammed in the fuel channel.

Gausden was given the bad news and quickly had to think of what to do next. He decided that they needed to remove fuel adjacent to the section that was on fire. That would form a firebreak and prevent the extreme heat from breaking open more fuel cartridges and spreading the fire. If more of the reactor caught fire, there was danger of overwhelming the stack filters. An eight-man team was formed, all equipped with airtight suits to prevent fission products from making body contact, respirators to prevent breathing of radioactive dust, and dosimeters to monitor their radiation exposures. They descended on the charge hoist and started pushing out fuel that was not burning.

Back in the general manager's office, the Windscale scientists gathered to figure out what to do. The problem was that the temperature had already risen to above the Wigner annealing point. There was, in theory,

AN EFFECT ON MILK PRODUCTION

The Windscale fire was the first nuclear power accident in which the general population was exposed to wind-borne fission products, and much was learned about the need for radiation dose standards from the aftermath of this incident. The ROF Sellafield site had been chosen partly for its proximity to the sea and the prevailing winds, with the thought that if any airborne radiation escaped from the facility, it would be blown harmlessly out into the ocean. The wind was not entirely cooperative, and significant radioactive debris from the fire, escaping the Cockcroft filters, made it to the adjacent farmland.

The most dangerous isotope produced by fission is probably iodine-131. What makes it hazardous is that the human body scavenges all the iodine it can from ingested material and concentrates it quickly in the thyroid gland. The efficient concentration of the isotope is what makes it a particular hazard. The most common carrier of iodine is milk, and the rural land around Windscale was home to a milk products industry. Iodine-131 dust falling out of the air onto the grass was consumed by cows all over Cumbria.

The health physics department at Windscale sent vans out into the countryside, specially equipped with radiation sampling and measurement equipment, and by Friday afternoon, October 11, just four days after the fire had started, they began to another Wigner energy release point at a higher temperature, thought to be at 2,192°F (1,200°C). If so, it could be a true disaster, with the temperature shooting up by thousands of degrees, destruction of the reactor, and the spread of fission products all over Great Britain. As the discussion continued, reports came in that the core temperature had reached 2,192°F (1,200°C), without an explosion. It was 5:00 P.M. The scientists worked to revise their predicted crisis temperature to 2,732°F (1,500°C), as Davey called Tuohy at home. Although Tuohy was at home caring for his sick family, this was an emergency, and Davey needed all available expertise. "Come at once," he said. "Pile number 1 is on fire."

Tuohy rushed straight to the reactor and took the elevator to the charge hoist, where men were desperately trying to remove fuel. The bottom half of the reactor front was a mass of flames. He retreated to the elevator, went to the top floor of the reactor building, and removed a view-port cover, so that he could inspect the back of the reactor. At 7:00 P.M., a red glow was

detect iodine-131 in milk from local dairies. The British government was careful not to cause a panic, and they issued no orders for an evacuation, releasing very little information about the accident at Windscale. The amount of radiation in the milk was between 0.4 and 0.8 microcuries per liter. At the time the limit of the threshold dose of radiation, below which there was no detectable effect, was 200 rad (2 Gy). Given this permissible dosage, it was determined that no milk more than 0.3 microcuries per liter should be served to infants. For the next month all milk in the area was tested, and most was confiscated, diluted with water, and then dumped into the Irish Sea, with affected dairy farmers compensated at the going rate for milk. The half-life of iodine-131 is eight days, so in one month the radiation level dropped by four half-lives, or to 0.0625 of the original dose rate. Today, no person, much less an infant, would be allowed anywhere near a 200 rad (2 Gy) dose of radiation to the thyroid gland. The current permissible radiation dose for the public from iodine-131 is approximately 0.5 rad (5 mGy) per year.

Follow-up studies of the 450 men who participated in fighting the Windscale fire in 1957 over 30 years revealed no unusual incidences of cancer or any radiation-specific illnesses. Statistical studies predict that as many as 200 Britons may have contracted cancer from low-level exposure to the fallout radiation from the fire. However, there is no clear-cut connection between low-level radiation exposure and cancer incidence, and statistical averaging is the best that can be reported on radiation-induced cancers.

visible. At 8:00 P.M., yellow flames were shooting out, and at 11:30 P.M., the flames had changed to hot blue. The fire was spreading. Tuohy fully understood the dangers of putting water on hot graphite, but there did not seem a choice. They could not just let the reactor go up in flame. He called the Windscale fire department and asked them to stand by with all available water pumping equipment.

By midnight, the crew was still working to contain the fire and had managed to keep it to 120 fuel channels. They tried to remove burning fuel, using every steel rod they could find, including scaffolding poles from the adjacent construction of the Calder Hall power reactor. Poles would come back out of the burning fuel channels glowing red and dripping with molten uranium.

Early morning Friday, October 11, Davey was feeling weak and dizzy from incipient flu, and his fellow managers persuaded him to go home. At 1:00 A.M., the local constable was awakened and advised of an emergency, with possible evacuation of the area. By 4:00 A.M., everyone agreed that it would be a disaster to put water on the fire, but they had to do something. A tanker truck of liquefied carbon dioxide was called for, and hoses were rigged to pipe it to the reactor and smother the fire.

Flushing the reactor with carbon dioxide had no visible effect on the fire. Tuohy went back to the inspection holes on top and tried to remove a cover plate. It seemed welded to the floor. The suction of the fire as it sought oxygen in the air was causing a low pressure in the reactor. With great effort he managed to lift the cover and peer into the inspection hole. Blue flames were now shooting out the back of the reactor and impacting the thick concrete wall 12 feet (3.7 m) away. The bio-shield was now under stress, and if it failed the entire Windscale facility would be lost. Anguishing over the need for a decision, he called for water to be poured into the fire.

It was not a simple matter to flood the reactor with water. The building had been designed to specifically have no access to water, so piping and distribution would have to be improvised. Fire-hose nozzles were tied to poles and shoved into the fuel channels. At about 9:00 A.M., the water was turned on. There was no explosion, nor did the fire rage hotter. Tuohy lifted a port cover and looked once again at the back of the reactor. There were still flames shooting out, but water was cascading down the back face. He had a further idea. The blowers had been kept on so that the men could work on the reactor face without dropping from the heat. He ordered all blowers stopped. The fire went out quickly. Just to make certain that it would not restart, the water was kept flowing for 30 hours, flooding the ground floor of the building with contaminated water.

A major emergency at the Windscale plutonium converter plant, as the reactor burns out of control *(© Mary Evans Picture Library/The Image Works)*

One of the Windscale graphite reactors remains at Sellafield, England—it is now inoperative. *(© Hulton-Deutsch Collections/Corbis)*

The fire had consumed 20 tons (18 tonnes) of uranium fuel, laced with fission products from having produced power. If Cockcroft's filters had not been installed at the top of the air stack, then northern England would have been blanketed with radioactive fallout. The term "Cockcroft's follies" was never used again. However, while the filters were very effective in keeping the bulk of the radioactive smoke inside the building, they did not stop everything. Around midnight on Thursday, before the firebreak was established and when the fire was fully engaged, significant radioactive dust escaped. Friday morning, when water was first used on the fire, steam carried radiation up the stack. The full extent of the release is difficult to measure, but a best estimate is that 20,000 curies (7.4×10^{14} Bq) of mixed radioactive material fell on the English countryside. In comparison, the similar release from the burning graphite pile at the Chernobyl disaster in the Ukraine in 1986 was 1,000 times greater. A single aboveground nuclear test in the United States, the Sedan test of 1962, dumped 44 times as much radioactive fallout on the midwestern states as the Windscale fire rained down on northern England. Still, it is among the worst nuclear accidents in history.

Windscale Unit 1 was sealed off and never repaired. Unit 2 was shut down and never restarted, and the British atomic bomb program had to devise other plans for obtaining plutonium-239. The Windscale reactors remain in place at Sellafield in Cumbria, England, currently owned by the Nuclear Decommissioning Authority. Robotic disassembly of the Windscale reactor cores is scheduled for completion by 2037.

Much work has been conducted to pin down the cause of the Windscale fire, without conclusive results. It is generally agreed that there are two possibilities: The fire was either due to improper annealing procedures or the failure of a special cartridge in the core, designated AM. The annealing was applied to the reactor before it was necessary and therefore too much heat was applied too rapidly. The first heating, on Monday, was unsuccessful, and on Tuesday the reactor was restarted in an attempt to bring it back up to annealing temperature. The reactor was more rapidly brought to full power. At this point, an aluminum-clad fuel cartridge probably broke open under the second application of high heat, and the fire started slowly. By Wednesday, the increased heating from the first smoldering cartridge set other cartridges on fire, and by Thursday a larger section of the reactor was burning.

The second possibility involves an alternate function of the Windscale piles. In addition to producing plutonium-239 for bombs, they were also used to produce *tritium,* for use in bomb initiators, or neutron generators using a deuterium-tritium fusion reaction. Tritium was produced in special cartridges containing lithium-magnesium rods, clad in aluminum to prevent fire. Named Mark III AM cartridges, these were loaded along with the uranium fuel in designated fuel channels. At temperatures more than 482°F (250°C), the lithium would start to diffuse through the aluminum cladding, taking about an hour to reach the surface and start burning. This would break open the can, and the resulting magnesium fire would be catastrophic, setting fire to adjacent uranium cartridges.

The Windscale fire shook the entire nuclear industry, bringing into question existing reactor designs, the use of graphite as a moderator, and a need for an impenetrable barrier between the fission process and the general population. In the nuclear power industry, there was no place for panic-mode engineering, rushed deadlines, or building a reactor before its design was completed. Another air-cooled power reactor was never built, and light water moderation became the dominant reactor type as graphite designs fell out of favor. Simple economy was no longer a nuclear industry design consideration.

A MAGNOX FUEL INCIDENT AT CHAPELCROSS

The experimental Calder Hall power reactors at Windscale were considered stable, safe graphite reactors, and the production group of the United Kingdom Atomic Energy Authority decided to replicate these reactors in the Dumfries region of southwest Scotland. Four Calder Hall reactors were built on the site of a World War II training airfield, RAF Annan, and construction was completed in 1959. The new *Chapelcross Nuclear Station* was officially opened for power production on May 2, 1959. It consisted of four 180-megawatt, graphite-moderated, carbon dioxide–cooled reactors using the same *MAGNOX* fuel as the Calder Hall plant. There was a second purpose for these MAGNOX reactors, making them essential for the British nuclear weapons program. In addition to making civilian electrical power, they were also used to make plutonium and tritium for use in bombs. Being military production reactors, they were restricted from examination by the international nuclear safeguards committees or watchdog groups.

MAGNOX cladding, which is an aluminum-magnesium alloy, is advantageous for reactors fueled with natural uranium. The probability of parasitic, nonproductive neutron capture in the cladding is very low. However, it also has two disadvantages. At high temperature, it catches fire, and it corrodes in water, preventing long-term storage in cooling pools. After having produced power in a reactor, the fuel cartridges have to cool off sitting in water before they can be chemically reprocessed, but there is a limit to how long they can stay in the coolant. The temperature restriction severely limits the power level of the reactor.

On May 19, 1967, a fuel cartridge overheated in a fuel channel in Reactor 2 at Chapelcross. It caught fire and melted, causing a great deal of concern. Analysis using a remote camera showed that graphite particles had detached from inside the fuel channel, piled up, and eventually blocked the flow of coolant gas. The Windscale fire was still fresh in engineering memory, and the prospect of a fire in another graphite reactor was unacceptable. It did not help the situation when the inspection camera broke off its mounting and fell into the reactor. The reactor was down until 1969, when successful clean-out of the fuel channel was completed. Three engineers were suited up and sent down to the reactor

British Gas-Cooled MAGNOX Reactor

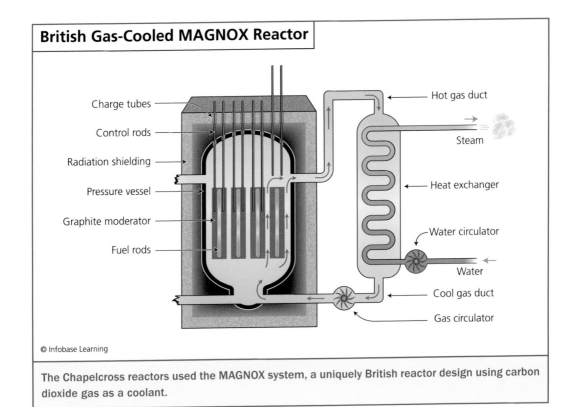

Charge tubes

Control rods

Radiation shielding

Pressure vessel

Graphite moderator

Fuel rods

Hot gas duct

Steam

Heat exchanger

Water circulator

Water

Cool gas duct

Gas circulator

© Infobase Learning

The Chapelcross reactors used the MAGNOX system, a uniquely British reactor design using carbon dioxide gas as a coolant.

face, with each given a three-minute window to scrub out the fuel channel with a long pole. A reason for the graphite pileup in the fuel channel, and therefore a lesson learned, was never firmly established.

Problems at Chapelcross seemed to peak in 2001. In the first quarter, an accident occurred while workers were refueling Reactor 2. The remote control machine, used to grab and hold a fuel element while it is being withdrawn from the reactor, snagged on the fuel and lifted it out of its shielding. This exposed the workers standing atop the reactor to a blast of intense radiation from the used fuel. This was classified a Level 1 incident, and it revealed shortfalls in the safety procedures and the equipment being used for fuel handling. Refueling was stopped for an accident investigation and a redesign of the procedures. Permanent changes were implemented to prevent the accident as the refueling operation continued.

In July 2001, rainwater leaked into some corroded steel drums containing uranium trioxide, leaching contamination out onto the ground.

This problem was discovered before things got out of control, and after an investigation the storage containers were upgraded to stainless steel drums, incapable of corrosion. The same month, three MAGNOX fuel cartridges slipped off a discharge chute, used to convey used fuel from the back of the reactor to the "acceptance bay," and fell 80 feet (24.4 m) into the water-filled cooling pond. MAGNOX cans are fragile, and breaking one open, exposing the burnable uranium to air, is to be avoided. In September, during the refueling outage, damage from Wigner growth was recorded in Reactor 1 and Reactor 4, and these units were never returned to operation. Power generation at the Chapelcross Works was officially ended in 2004, and consent to begin decommissioning of the power plant was granted on September 29, 2005.

There were 11 MAGNOX gas-cooled, graphite reactor power plants built in the United Kingdom. The last unit is scheduled to be shut down in 2011. MAGNOX plants were also exported to Italy and Japan, and both reactors have been shut down. North Korean engineers developed their

The Chapelcross Nuclear Station near Annan, Scotland, being dismantled (© *Global Warming Images/Alamy*)

own copy of a MAGNOX reactor, based on available blueprints of the Calder Hall reactors. Construction of the power station at Yongbyon was ceased in 1994, along with another, more powerful MAGNOX reactor under construction at Taechon, under the U.S.-North Korea Agreed Framework for nuclear disarmament. By 2004, the parts that had been built, unprotected from the weather, had deteriorated badly, and in 2010 the reactors were torn down. The MAGNOX fuel-processing facilities, built along with the reactors, were disabled in 2007 in accordance with international agreements, but they were reactivated in 2009 for the purpose of producing *atomic bomb* material.

The Windscale facility has been accused of dumping diluted radioactive waste into the Irish Sea, particularly during the late 1940s and 1950s, possibly making this section of the North Atlantic one of the most contaminated seas in the world. This situation is being monitored and measured by the Convention for the Protection of the Marine Environment of the North-East Atlantic (the OSPAR Convention). OSPAR reports an estimated 440 pounds (200 kg) of plutonium-239 in the bottom sediments of the Irish Sea.

Since April 1, 2005, the Windscale facility is being operated by the private company Sellafield Ltd and is owned by the Nuclear Decommissioning Authority. Efforts are underway to carefully decontaminate and decommission all of the outdated nuclear facilities at Windscale. There is no hurry, and these operations will continue for at least another 40 years.

The next chapter covers some technical anomalies at the top of the ladder in experimental power reactors. These operations were far from the normal routine reactor experiments and include special reactors and radioisotope power systems built for use in satellites, a small-scale liquid metal–cooled fast breeder reactor, and a large-scale version of the same design. The hazards of these systems were unique, but useful lessons were usually learned from the inevitable accidents.

5 Highly Technical Problems

A nuclear power accident is defined as an unanticipated or unscheduled occurrence in a nuclear power setting. It is not necessary for the incident to have a direct health effect on a human being. The simple fact that some aspect of an operation or an experiment has not proceeded exactly as planned, with its proximity to a potentially dangerous nuclear process, is sufficient to cause concern. The fact that in-place safety systems have correctly responded to a developing situation is not acceptable if those systems have reached a limit of effectiveness with no remaining margin. As a rule, reportable, investigated incidents involve threats or potential threats to the safety of workers or the general public.

This chapter covers some purely technical problems, including the *Fermi-1* meltdown in 1966. This reactor was the first attempt to build a full-scale liquid metal–cooled fast breeder reactor in the United States, using the most sophisticated nuclear technology available. This chapter also looks at a serious accident inside the small world of experimental nuclear engineering, at the *Santa Susana Field Laboratory*. The sidebar examines a nonthreatening but real nuclear reactor accident in outer space.

EARLY TROUBLE AT THE SANTA SUSANA FIELD LABORATORY

The Santa Susana Field Laboratory (SSFL) was established in 1947 by North American Aviation as a development and static-fire test station for rocket engines. The site chosen for the 2,850-acre (11 km²) laboratory was on a hill overlooking the Simi Valley to the north, about 30 miles (48 km) from downtown Los Angeles, California. In 1953, the test and development activity was expanded to include nuclear research in Area IV, apart from the rocket development group, which operated in Areas I, II, and III. In December 1955, the two research groups became separate divisions, with Atomics International responsible for the nuclear work and

The Santa Susana Field Laboratory, Area IV, in 1962 *(Boeing Company)*

Rocketdyne in charge of the rocket propulsion research. Rocketdyne developed the Redstone short-range missile, the Thor intermediate-range missile, the NAVAHO intercontinental cruise missile, and engines for the Apollo Moon-landing missions.

The Atomics International facility was an experimental nuclear laboratory, always building and testing new hardware. Accidents and radiation spills were not uncommon in the early years of active testing in the nuclear research laboratories, which were purposefully located in remote, underpopulated areas. The worst nuclear accident at Santa Susana involved the Sodium Reactor Experiment (SRE). When the SRE project started, in 1954, there was concern that water-moderated and cooled reactors were subject to steam explosion if the core overheated. To make such an accident impossible, the SRE was graphite moderated with sodium metal used as the coolant. The sodium, liquefied by the heat of the reactor, was pumped through the uranium-fueled core in a closed loop. No matter how hot it became, the liquid metal could not flash explosively into steam, and the solid graphite provided neutron moderation.

The SRE was a full-scale experimental power reactor, and it included a turbogenerator that could be switched into the California power grid. After three years of construction and testing, the SRE was started up in early July 1957. On July 12, 1957, it was connected to the power lines, feeding electricity to 1,100 homes in the City of Moorpark. It beat the Shippingport reactor in Pennsylvania by a few months in becoming the first commercial nuclear power station in the United States. The pressurized water reactor at Shippingport started generating electricity in December 1957.

The experimental reactor performed well until July 13, 1959, when the power level suddenly started increasing for no apparent reason. With a great deal of effort, the operations crew managed to bring it under control and eventually shut it down. In a few hours of puzzling over data collected during the power excursion, the operators, engineers, and scientists of Santa Susana could not find what had caused this alarming behavior. They then made the most grievous of reactor operations errors: Without knowing why the reactor had experienced runaway power, they decided to start it up and see if it would do it again.

The reactor started and came up to operating power, but it never seemed to be operating correctly. On July 26, 13 days after the power incident, high radiation alarms started going off, and at the end of the month the design team decided to shut it down and find out what was wrong. It is not easy to examine a sodium-cooled reactor for mechanical defects, because

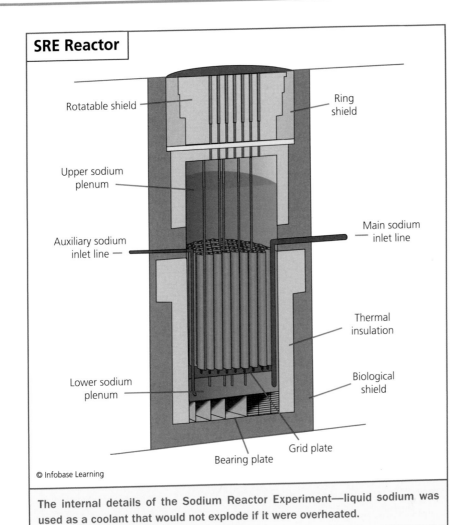

SRE Reactor

Rotatable shield

Ring shield

Upper sodium plenum

Main sodium inlet line

Auxiliary sodium inlet line

Thermal insulation

Lower sodium plenum

Biological shield

Bearing plate

Grid plate

© Infobase Learning

The internal details of the Sodium Reactor Experiment—liquid sodium was used as a coolant that would not explode if it were overheated.

the reactor tank is filled with radioactive, opaque sodium, and it must be drained carefully or the heat production remaining in the fuel can cause problems. They lowered a camera into the core and found that 13 of the 43 high-temperature uranium-carbide fuel elements, or almost a third of the core, had melted. With that finding, the entire system was dismantled to discover why the fuel had melted.

An organic substance, tetralin, had been used as a coolant for the seals in the liquid sodium pumps. The seals had leaked, tetralin seeped into the flowing liquid sodium, and it carbonized, making a solid material. The burned tetralin started to build up in the cooling channels between the fuel rods. Once a small amount of carbon stuck to the cooling channel walls, more would hit it and stick. Eventually, entire cooling channels were blocked with carbon, and the coolant failed to flow. Without coolant, the fuel melted, and this changed the configuration of the reactor core, making it unpredictable and difficult to control.

Unfortunately, the SRE was not housed in a containment building, as now is strictly required for all commercial nuclear reactors to prevent any radioactive contamination of the environment surrounding the facility. It was housed in a standard industrial building, which was never intended to prevent the leakage of radioactive gases into the

A NUCLEAR ACCIDENT IN OUTER SPACE

The SNAP-8 was a nuclear fission reactor designed by North American Aviation at the Santa Susana Field Laboratory in California. The purpose of this small reactor was to deliver high electrical power on demand in an orbiting spacecraft. It could generate 35 to 90 kilowatts, making it capable of supporting satellite missions with significant electrical power loads, such as ground-scanning radar. The fuel was highly enriched uranium-235, and the coolant was a mixture of sodium and potassium metals. The coolant transferred heat to liquid mercury, which would boil, and thus power a small turbogenerator. The fuel pins in the SNAP-8 were delicately thin, and an experiment with one of two prototypes experienced a meltdown.

Lessons learned from the SNAP-8 experiments resulted in the design of the first nuclear reactor to be boosted into space, the SNAP-10A. The power generation was simplified to solid-state *thermoelectric* converters, heated by the sodium-potassium coolant, and the electrical power output was reduced to 500 watts. The core was designed around 37 fuel rods, each made of a uranium-zirconium-hydride material, at the center of a compact cylindrical reactor 16 inches (40 cm) long and nine inches (22 cm) in diameter. The fission reaction was controlled by four half-cylindrical drums, made of beryllium, designed to be rotated in and out of proximity to the reactor core, varying the amount of escaped neutron flux reflected back into the reactor. With the drums rotated to face the reactor core, neutrons streaming out would be directed back into the core, increasing the level of fission. The drums were moved with electric motors and were held in place by a retaining band anchored to the reactor chassis by an explosive bolt. Steel springs were compressed between the drum axles and environment. Dangerously radioactive gases, including iodine-131, were piped from the reactor building into holding tanks and then bled into the atmosphere at a controlled rate over a period of several weeks. Although the fuel damage was immediately announced with a press release, the full implications of this event were not fully understood by the public or by the nuclear engineering community until decades later.

There were no reported injuries, but the full extent of the radiological contamination caused by it is unknown. The reactor was rebuilt, restarted, and continued operation until 1964 without further incident or radiation leakage.

the chassis. In the event of an emergency, a scram signal would fire the explosive bolt, sending the four control cylinders flying away from the reactor and shutting it down instantly. A scram would be irrecoverable, as there would be no way to collect and reassemble the far-flung controls. The reactor was equipped with no radiation shielding, as it would be started and run in orbit, without a living thing in range of the extreme radiation flux caused by fission and fission products. The entire SNAP-10A weighed only 650 pounds (295 kg).

After extensive ground testing at Santa Susana, the SNAP-10A was boosted into orbit aboard an ATLAS Agena D rocket on April 3, 1965, from Vandenberg Air Force Base. Code name for the project was SNAPSHOT. The reactor was started with a command from the ground control station, and telemetry signals from the satellite indicated that the equipment was performing as designed, producing 500 watts of electricity.

All was well for 43 days, when the telemetry signals suddenly stopped. With some effort, the telemetry link was restored, and the ground controllers discovered that a relay, used in a voltage regulator circuit, had burned out. This glitch caused the system to scram, and the controls were kicked out of the reactor. There was no way to restart it. It was the first mechanical failure of a nuclear power reactor that could not possibly affect a human being. To ensure the continued safety of everyone on Earth, the Agena D rocket was restarted, pushing the now dead reactor into a 700-nautical-mile (1,300-km) orbit. This orbit should be stable for the next 4,000 years. The United States never put another nuclear fission reactor into space. The four control cylinders are tied to the reactor with steel tether cables and are not floating separately as space debris.

As part of cleanup operations at the Santa Susana Field Laboratory, buried radioactive waste is excavated for relocation. *(DOE Photo)*

From 1957 until 1988, the nuclear research division at Santa Susana operated the largest "hot lab" in the United States, used to chop spent uranium fuel rods into small pieces. Fuel from all other Atomic Energy Commission (AEC) and, later, Department of Energy *(DOE)* nuclear projects was shipped to Santa Susana for de-cladding and examination. Used reactor fuel is extremely radioactive from a buildup of fission products, and all handling must be by remote control. In 1957, a fire broke out in the hot lab. The fire got out of control, and massive contamination of the building resulted. Another fire broke out in the building in 1971, involving sodium-potassium coolant mixed with fission products.

It is difficult to dismantle a sodium-cooled reactor and dispose of the coolant. Liquid sodium becomes highly radioactive after service in a reactor, and it explodes when exposed to water. Even slight water vapor in the air causes a severe reaction, and the smoke carries away radioactive debris. At Santa Susana, old sodium coolant and sodium-contaminated parts were burned in an open pit. Another disposal practice was to shoot a barrel filled with potentially toxic waste with a rifle at a safe distance, so that it would explode and release the contents into the air. Although there was nothing illegal about these procedures at the time, it is doubtful that such methods of disposal will ever be used again. The cleanup of the Santa Susana Field Laboratory, which has remaining traces of radiological contamination from more than 30 years of nuclear research, continues to this day. The SSFL is now jointly owned by NASA and the Boeing Company.

A CORE MELT IN THE FERMI 1 FAST BREEDER REACTOR

Walker Lee Cisler (1897–1994) was a mechanical engineer with a long and distinguished career in electrical power production when he became chief engineer at Detroit Edison in 1943. In 1947, he was named executive secretary to the AEC's Industrial Advisory Group, and the commercial application of nuclear fission became his life's goal. By 1951, encouraged by work on the EBR-1 experimental breeder reactor at the National Reactor Test Station, Cisler was advocating the construction of a commercial-scale fast breeder reactor in Michigan, to be operated by Detroit Edison to provide inexpensive power for the city of Detroit. At the time, nuclear power had great prospects as a cheap, pollution-free energy source, but uranium fuel reserves had not been established. A breeder reactor power station would never want for fuel.

Designs were drawn for a breeder reactor with commercial power plant safety measures, including a full containment building to prevent the introduction of radioactive materials into the environment in case of total breakdown of the reactor. Two companies, the Power Reactor Development Company and the Atomic Power Development Associates, combined resources to fund the construction, and in January 1956 Cisler applied for the permit. Construction cost estimates grew to $70 million, as the AEC granted $4 million for design research and special equipment. Despite vigorous opposition from Walter Reuther (1907–70), a labor union leader, and the AFL-CIO, the construction permit was granted on May 26, 1959, and the Fermi 1 Nuclear Generating Station was built on the shore of Lake Erie, at Lagoona Beach, Michigan.

The nuclear fission reactor used enriched uranium-235 fuel, breeding plutonium-239 in a surrounding blanket of uranium-238 and generating

94 megawatts of electricity. Once a full core load of plutonium-239 had been converted in the blanket, the reactor could be switched to using the plutonium for fuel, and the breeder-reactor cycle would be complete. The coolant had to be a heavy substance to prevent neutron moderation in the core. Only "fast" neutrons were usable for breeding plutonium, so only fission at the high end of the neutron energy spectrum was possible. Pure sodium metal, made liquid by the heat of the reactor, was chosen as the coolant. It would be melted in the reactor core and pumped into a heat-exchanger, where the hot sodium boiled water for a turbogenerator. There are problems with using sodium as a coolant, and all would plague the Fermi 1 project.

The first sodium accident occurred on August 24, 1959, when a quantity of sodium was allowed exposure to air as a test of fire-control procedures. The test did not proceed as planned, and there was a sodium explosion. Six people were hospitalized, many more were injured, and several homes in the nearby communities of Trenton and Riverwood were damaged. A full-scale sodium-cooled reactor had never been built before, and there was much to learn about its characteristics under high-power conditions. In 1960, experiments with sodium indicated that it would react unfavorably with the graphite neutron reflector surrounding the reactor, and a major redesign was called for, costing $2.5 million and 15 months of construction delay. On December 12, 1963, a test of the sodium circulation system was conducted. The reactor had yet to be loaded with fuel, but the sodium coolant was installed and the pumps were operational. It was noted that there was an odd "clapping" sound in the sodium loop, possibly originating in the reactor tank. As the reactor operators sat at the console and monitored the instruments, the temperature in the cooling loop suddenly rose. An operator hit the emergency dump switch for the water in the secondary steam loop, thinking that water had leaked into the sodium loop, but it was too late. A pressure-relief valve blew open, and raw sodium streamed out into the building. If the reactor had been operating with a fuel load, this accident would have been severe, as the sodium would have been radioactive. Hearings with the AEC were held, operating procedures were rewritten, and emergency preparations were strengthened. Preparations for power generation continued.

The Fermi 1 breeder reactor had been well designed, with safety measures considered in all aspects of the system. Underneath the core, on the bottom of the reactor tank, was mounted a cone made of steel. The purpose of this odd structure was to ensure that if the core melted, then the liquefied uranium would be forced to spread out at the bottom of the tank and not clump together in the middle. This would prevent the melted core from assuming a critical configuration in which the reactor would still be generating power after the emergency of a meltdown. This was a thoughtful design, and if it had remained strictly as drawn in the plans it would have been fine. The builders, however, committed a serious error. Thinking that the design could be improved by covering the steel cone with zirconium, which was known for good mechanical properties under extreme heat and radiation, they superseded the design by fabricating

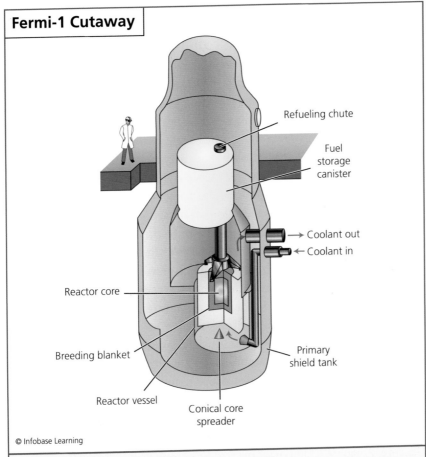

Fermi-1 Cutaway

Refueling chute

Fuel storage canister

Coolant out

Coolant in

Reactor core

Breeding blanket

Primary shield tank

Reactor vessel

Conical core spreader

© Infobase Learning

A cutaway diagram showing the internal workings of the Fermi 1 fast breeder reactor. The cone-shaped core spreader, added as an afterthought during construction, would cause an expensive accident.

a zirconium skin for the deflector cone on site. This design change was never noted on the plans, but the builders considered it an improvement over what had been drawn.

At 12:30 P.M. on August 23, 1963, the Fermi 1 Nuclear Generating Station first achieved sustained nuclear fission. The power level was kept low for several months, as bugs were worked out of this new, unique power plant. By January 1966, the power level had been increased to 80 megawatts. On October 5, 1966, Fermi 1 was poised to break another record for the highest power produced by a breeder reactor, and the operating crew was prepared to push it to the next step. The power level was increased beginning at 8:00 A.M., and almost immediately there was a problem. A valve malfunctioned in one of the steam heat-exchangers, and it took until 2:00 P.M. to clear the problem. By 3:00 P.M., the power level was up to 20 percent of the goal.

Five minutes later, problems began to indicate on the control console. First, the control rods had been pulled out too far to achieve criticality. Six inches (15 cm) should have been enough pull, but they were out nine inches (23 cm). The neutron counter readings, used to indicate criticality of the fission, were jumping in an unnatural fashion, as if the instruments were plagued with electrical noise. Fuel assembly number M-140 began reading a temperature of 700° Fahrenheit (371°C). The fuel assemblies were supposed to operate at 580° Fahrenheit (340°C).

At exactly 3:09 P.M., the radiation warning air-horn began sounding, indicating that fission products were loose in the building. All doors and windows were sealed, and the air-vents were closed. Hoping to avoid the thermal and mechanical shock of a full scram, the operators began to lower the power level by easing the control rods in, slowly. By 3:20 P.M., it was decided to scram the reactor, and an operator hit the big red button. All the control rods slammed into shutdown position except one. It was stuck six inches (15 cm) from the fully down position. Scram was initiated again, and this time the rod fell fully in. The stuck rod was ominous. The only thing that would cause a control rod to hang up was a bent fuel assembly, and the only thing that would bend fuel was a melting of the uranium.

It was obvious that fuel had melted, but the reason it had reached melting temperature was not at all clear, nor was the extent of the damage in the core. A sophisticated machine had been built and installed to automatically load and unload the fuel from the reactor core, but it was not designed to handle melted fuel, nor was there any way to see into the core and simply look at the fuel assemblies. The core was under opaque sodium metal, and it was encased in a solid steel tank. The reactor had been designed to operate considering a fuel melt to be impossible. With engineers and operating crews working 24 hours a day, it took until January 1967 to discover the extent of the damage and until May 1967 to remove the damaged fuel. Although there was no provision in the design for draining out the coolant and no provision for storing it, the coolant level had to be lowered so the fuel could be seen. A viewing port was constructed for the top of the reactor tank, and, at high cost, a special borescope was built to be lowered into the core and photograph the fuel assemblies.

Six fuel assemblies out of 120 had melted, with two of them stuck together. It seemed a small error, but the smallest mechanical distortion of the fuel was enough to render the expensive power plant unusable. By August 1967, enough sodium had been drained to see the bottom of the reactor tank, and the borescope and light were lowered down below the bottom of the core. There they found what had been making a clapping noise in the coolant and what had stopped up four coolant channels at the bottom of the core. It appeared to be a crushed beer can, but they knew positively that no beer can had been allowed anywhere near the reactor as it was being constructed. The nature of this out-of-place object was a mystery. More special equipment was built, and an access hole was bored in a coolant pipe. A manipulator was built for the single purpose of flipping the beer can over so that they could see and photograph the bottom of it, and it had to operate through a distance of 40 feet (12 m). Moving the beer can from the top of the reactor was like performing an appendectomy through a nostril. The bottom of the metal object looked about like the top, and the engineers came no closer to figuring out what it was doing in the reactor tank.

Finally, a year and a half after the accident, another specially built tool managed to pick up the apparent beer can and lift it out the top of the reactor. It took an hour and a half just to lift it, pausing to let it cool and reach room temperature. Once they could examine it up close, the engineers had no trouble identifying the piece of metal. It was the zirconium cover that had been made for the steel spreader cone at the bottom of the reactor. It was not on the reactor blueprints, because it had been added at the last minute as an improvement of the design. It had popped loose in the severe action of the coolant stream and was sucked into the core coolant channels, where it blocked sodium from the hot fuel.

In May 1970, the Fermi 1 Nuclear Generating Station had been repaired and was ready for restart, but another sodium explosion caused a delay. Finally, in July 1970, the reactor was restarted, and it achieved a power level greater than its designed output, 200 megawatts, in October. Fermi 1 was shut down on September 22, 1972. Its request for an extension to the operating license was denied, and the plant was officially decommissioned on December 31, 1975.

The Clinch River Breeder Reactor project was begun with federal funding in 1971 to build a 1,000-megawatt power plant in Roane County, Tennessee, near Oak Ridge, based on lessons learned from the experimental reactors built in Idaho and the Fermi 1 in Michigan. In 1977, U.S. president Jimmy Carter denied an operating license to the Barnwell, South Carolina, fuel reprocessing plant on the grounds of possible terrorist infiltration and theft of plutonium. Without fuel reprocessing, by which plutonium-239 can be extracted from a mixture with uranium-238, a breeder reactor is pointless. The project was shut down, and no further breeder reactor construction has been attempted in the United States.

6 Accidents at Fuel-Processing Facilities

There are two purposes for nuclear fuel processing: Uranium is extracted from mined ore and formed into nuclear reactor fuel or spent fuel is taken from a nuclear reactor and plutonium is chemically extracted from it. There is no plutonium in unused reactor fuel, but as the uranium-235 is fissioned away in the power-production process, plutonium is created by the capture of neutrons in the inert component of the fuel uranium-238. Plutonium has little current commercial application, but it has been a key component in nuclear weapons since its first use in World War II. Many fuel-processing plants were thus built as military installations, and plutonium extraction is primarily a function of weapons production.

After World War II, the Soviet Union developed a military need for plutonium-239, and production plants were built in *Electrostal, Tomsk,* and the third largest city in Russia, *Novosibirsk.* The goal of these plants was to match the production volume of those in the United States. The UK nuclear installation at Windscale in Cumbria, England, also contained fuel-processing facilities for military applications. Japan, which has no military application for plutonium, nevertheless has a fuel-processing facility at *Tokai*-muro, supporting an experimental breeder reactor project and general fuel production for commercial nuclear plants.

The amount of material required to make a critical reaction varies according to the enrichment of the uranium or the density of the plutonium in the solution, but it usually goes critical in a volume that would fit in a trash can. The shape of the volume of solution can also make a lot of difference. The ideal shape is a sphere, followed by a cylinder with the exact proportions of a soup can, having a minimum surface-area-to-volume ratio, minimizing the escape of neutrons. A soup can is, after all, designed to be built using the smallest possible amount of steel and has this optimized ratio. A nuclear fuel process must be designed to allow no collection of fissile fuel in solution to reach or exceed the critical configuration. It is possible to have a large volume of fissile solution in one place in the plant, but the shape of the container must be made to have a large surface-area-to-volume ratio to avoid causing a nuclear reaction. A large surface area promotes neutron leakage and therefore prevents criticality, even if there is enough fissile material present to make a reactor. The safest design philosophy is to simply keep all volumes of solution from reaching a critical size.

If this rule is followed strictly, then a nuclear fuel process should be as safe as any other chemical operation. The simplicity of the rule would seem to make it foolproof. Instead, it makes each instance of a criticality accident in a fuel-processing plant unforeseeably complex. There have been 22 criticality accidents in nuclear fuel processing in the United States, Russia, Great Britain, and Japan. The following statistical characteristics of these accidents are interesting:

- ❈ Of the 22 accidents, 21 occurred with the fissile material in a solution or a slurry.
- ❈ One of the criticality accidents occurred with a pile of metal ingots.
- ❈ No accident has ever occurred with powders of fissile material.
- ❈ Of the 22 accidents, 18 of them occurred in unshielded, manned facilities.
- ❈ There were nine fatalities.
- ❈ Of the survivors, three had limbs amputated.
- ❈ No accident has ever occurred in the transportation of fissile material.
- ❈ No accident has ever occurred while fissile material was being stored.
- ❈ Only one accident resulted in significant radiation exposure to the public.
- ❈ Only one accident resulted in the buildup of measurable fission product contamination.

Only through study of these characteristics has the processing of nuclear fuel improved and accidents been minimized. Between 1957 and 1971, there was a clump of 17 accidents, but since 1979 there have been only two reported criticality accidents.

FATALITY AND POWER EXCURSIONS AT THE MAYAK PRODUCTION ASSOCIATION

The *Mayak* industrial plant was built between 1945 and 1948 in the wilderness between the Russian towns of Kasli and Kyshtym, 45 miles (72 km) northwest of Chelyabinsk. The original mission of this plant was to extract, refine, and machine plutonium for nuclear weapons. In the 45 years of plutonium production, nearly half a million people were irradiated by fallout from criticality accidents occurring in the plant, exposing them to more than 20 times the radiation released by the Chernobyl reactor explosion in the Ukraine in 1986.

The most significant noncriticality accident at Mayak, not involving nuclear fission, occurred on September 29, 1957, when the failure of a cooling system for a tank containing tens of thousands of tons of nuclear waste dissolved in water caused the mixture to overheat and explode. The explosion had the estimated force of 75 tons (310 gj) of TNT, and it released 20 million curies (7.4×10^{17} bq) of mixed radiation. An estimated 200 people died immediately of radiation poisoning, 470,000 received injurious radiation exposures, and 10,000 were evacuated. This accident is classified as a Level 6 on the International Nuclear Event Scale (INES).

Russians driving through the area in the 1960s found a deserted region, where signs warned them to keep moving and not to roll down the windows. Knowledge of this incident outside the Soviet Union was minimal and was not confirmed and detailed until 1992, when the new Russian government acknowledged that the accident had occurred.

A serious criticality accident at Mayak occurred on April 21, 1957, in the uranium processing line. Several glove boxes were lined up on a wall, transferring uranium in solution from glove box to glove box using vacuum lines. A glove box is a sealed container with a glass window in front. It is used for working with chemicals without the danger of skin exposure. The worker puts his or her hands into long rubber gloves, built into the front of the box and sealed all around, allowing sensitive manipulation of anything in the box without actually touching anything with bare hands.

Monitoring radiation levels in the environment in Russia *(Thomas Nilsen/SPL/ Photo Researchers, Inc.)*

In this case, the worker was combining a small, strictly controlled volume of uranyl nitrate mixed with oxalic acid in a precipitation vessel, stirred with an electric mixer and temperature-controlled by a steam jacket around the vessel. The result of the chemical reaction in the precipitation vessel was a uranyl oxalate slurry, which was then drawn into a receiving vessel after being sucked through a filter. Criticality control was strictly through the limitation of the size of each batch, which was held to 28 ounces (800 gm) of uranium. The sizes and shapes of the vessels in the glove box were not adequate controls. There was no criticality alarm on the glove box, and the worker had no training as to the nature or the possibility of a criticality incident.

While doing a perfectly routine filtration of uranyl oxalate out of the precipitation vessel and into the receiving vessel, the worker, peering through the glass window, noticed that the filter disk seemed to be bulging upward in its transparent holder. If he could have seen into the opaque receiving vessel, he would have seen the uranyl oxalate suddenly boiling furiously. Before he had a chance to wonder what was going on, a violent release of gas from the receiving vessel blew out the filter and precipitate onto the floor of his glove box. Instinctively, the worker began scooping up the spilled precipitate to load it back into the filter housing, but within a few seconds he became deathly ill. After about 10 minutes, enough material had been ejected from the receiving vessel to stop the supercritical uranium-water mass in the vessel from reacting.

A radiation control person was called in. He measured the radiation steaming from the glove box and ordered an immediate evacuation of the building. The worker had received a dose of 3,000 rads (30 Gy). He died 12 days later. Five other workers had been in the vicinity, and each had received about 300 rads (3 Gy) of gamma radiation. All suffered temporary radiation sickness, but they recovered without apparent long-term injuries.

Because the accident had happened inside a glove box, the cleanup was not difficult and operation resumed within a few days, but the cause of the accident had to be determined, and it was not obvious. A critical mass of uranium in water had collected in the receiving vessel, even though the worker had followed all procedures correctly. In the cleanup operation, it was discovered that a thin, hard crust of precipitate had built up in the receiving vessel. As this mass of uranium was not inventoried as existing material, the additional uranium in solution added to the vessel had resulted in a mass of 6.75 pounds (3.06 kg) of uranium in the vessel, which was more than enough to cause criticality. A procedure change from two months before, streamlining the flushing of the vessels using a simple wash of nitric acid, is thought to have been the cause of the accident.

A radiation meter was installed above the glove box, operating instructions were revised, and the workers received criticality safety training. As a result of the accident, an experiment to better determine the critical factors for the uranium purification process was implemented. Unfortunately, the next criticality accident was due to this experiment.

After the two criticality accidents at Mayak, a team of four scientists set out to determine how this could be prevented in future operations. They set up a carefully designed experiment to determine the volumes and shapes of different concentrations of uranium and plutonium in solution that were capable of going critical during the fuel processing. An open stainless steel tank was bolted to a steel stand, about 2.6 feet (0.8 m) off the concrete floor and separated from the walls of the room. The tank simulated a typical fuel-processing vessel, and various concentrations of fissile materials were slowly dripped into it in standard "approach to criticality" experiments. It could hold up to 106 gallons (400 l) of fuel solution. Neutron and gamma ray counters continuously monitored the activity of the contents of the tank during an experiment. Criticality was approached but never achieved during an experiment. The distance from the floor and walls was important, as it simulated actual plant conditions. The concrete or anything else near the experimental tank would reflect neutrons back into the solution and improve the reactivity of the mass of fuel inside. When criticality was approached, simply waving a hand over the tank could reflect enough neutrons back into it to throw it over the top, into a critical reactor situation. The contents of the experimental tank were always emptied afterward into multiple 1.6-gallon (6-l) containers, too small to ever contain a critical mass, using vacuum lines.

The team of four was dedicated and well educated. On January 2, 1958, it was the first working day of the year, immediately following the New Year's holiday, and they were the first shift, starting work at 1:00 A.M. The first experiment of the shift went exactly as intended, with a large volume of fuel solution left in the experimental tank, barely below criticality. The team proceeded to empty it into the small containers. Growing weary of the tedium involved with slowly draining the tank into the little bottles, the team decided that the solution remaining in the tank was far subcritical by now, and they could speed up the process. They unbolted the tank from the stand so that they could tilt it and pour the remaining contents into the bottles. It was heavy, but three of them managed to slide it halfway off the stand and tilt it.

Instantly, a blue flash lit up the room. The fissile solution remaining in the tank boiled violently and shot out the top of the tank, hitting the ceiling from a distance of 16 feet (5 m). They dropped the tank on the floor and ran to the change room. Knowing full well that they had caused a criticality, they tried to shower off the radioactive fission products with which they were all soaked. The crew of four was then transported to the plant hospital. The three who had lifted the tank were each dosed with as much as 8,000 rads (80 Gy) of radiation. In six days, they had all died. The fourth scientist, who had been standing off to the side, received a 600-rad (6-Gy) dose. She developed cataracts and lost sight in both eyes. The criticality experiment program was terminated immediately, and the apparatus was disassembled.

The cause of this accident was violation of the experiment procedures. They were supposed to drain all the fluid into separate bottles through the vacuum hose, regardless of how tedious the procedure seemed. Standing around the tank, the experimenters themselves contributed reflection to the setup, and lowering the tank to the floor encouraged neutron reflection off the concrete. Furthermore, when they tilted the tank, the remaining solution changed shape entirely. Instead of the inefficient shape of a shallow pool in the bottom of the tank, with a great deal of surface area from which neutrons could leak without producing fission, the shape of the fluid was cornered, becoming much more efficient. While it was true that the remaining solution was far from critical as it sat at the bottom of the tank, changing the shape and the amount of reflection can alter the situation in severe ways.

Mayak seemed to keep repeating the same accident at different points in the chain of the fuel process. On September 7, 1962, the facility experienced a chain of supercritical pulses in a tank in a glove box. This was in a building at the end of the process line where plutonium feed material was cast into metal. There was a slight amount of wastage in casting ingots of plutonium, as small contaminations would collect in sand, slag, and briquette chips. Plutonium was too precious to waste in any quantity, so these residues were dissolved and fed back into the process. Detritus containing plutonium residue was mixed with nitric acid in stainless steel vessels. The vessels were of adequate size and of a geometry favorable to fission, so it was only a matter of time until too much plutonium dissolved in a water-based solution would collect. The mixture pulsed with supercriticality three times before the operators were able to transfer enough liquid out of the vessel to bring it under control. No one was injured, but the fuel process was still not safe.

Yet another pulsating criticality accident happened in a glove box on December 16, 1965. This time it was in the residue recovery section of the metal and fissile solution processing building, in a dissolving vessel. These vessels held 26 gallons of liquid (100 l) and were of a perfect shape, cylindrical with a half-sphere bottom, to encourage criticality. Due to poor communication by telephone, the uranium concentration of batch number 1826, 0.32 percent by weight, was confused with that of batch number 1726, which was 44 percent by weight, or 138 times higher. The dissolving vessel was filled with batch number 1726, assuming it had the uranium concentration of batch number 1826, and the result was a short blast of the local criticality alarm. The operator wisely left his station, even though the alarm had only sounded once, and eventually several dozen alarms started going off all over the building. The criticality pulsed 11 times before the engineers were finally able to quench the reaction in the dissolving vessel by removing the gloves on the box for deep access, wadding up a cadmium foil, and dropping it into the vessel. The cadmium dissolved in the nitric acid, making a characteristic orange smoke, and then absorbed enough neutrons from the reaction to shut down the fission. No injuries were reported, but it was still dangerous to work at Mayak in 1965, 20 years after it was built.

Finally, a particularly tragic and unnecessary criticality accident occurred at Mayak on December 10, 1968. In the basement of the plutonium metallurgical building, a small group of workers had set up a small-scale research and development operation to find the purification properties of various organic solvents. The experiments were improvised, unauthorized, against all rules and regulations, and destined for failure. The researchers had two 264-gallon (1,000-l) stainless steel tanks bolted to the floor, and a 16-gallon (60-l) bucket sitting on a shelf above the tanks. The floor was concrete with a floor drain. Due to a confusing change of piping to the tanks, a small volume of organic plutonium solution had inadvertently collected atop a larger volume of aqueous plutonium solution in tank number 2. Tank number 1 was empty. The concentration of plutonium in the aqueous solution was much too weak to form a critical mass. The organic plutonium solution was very concentrated, but it was in a thin layer atop the weakly concentrated water. Although there was enough plutonium in the organic solvent to make a critical mass, the geometry, a thin disc, was extremely disadvantageous for fission, and tank number 2 was full but in a safe condition.

The shift supervisor wanted the organic contamination removed from the top of the contents of tank 2, and he set up an operator to suck the liquid off the top of tank 2 into a five-gallon (20-l) glass bottle, using rubber hoses and a connection to a vacuum line. The shift supervisor then left to check on other operations in the building. The operator managed to completely fill the glass bottle with solvent solution. He looked for somewhere to put it, found the bucket sitting on the shelf, and emptied the bottle into it.

He then hooked the hose up again and once again filled up the entire bottle. There was more solvent contamination in the tank than they had realized. He disconnected the hoses, hoisted the bottle up, and started to empty it into the bucket. The bucket was a very favorable cylindrical geometry, and the plutonium solution configuration was changed from a thin disc in tank 2 to a taller cylinder in the bucket. The operator saw a flash of blue light and felt a wave of extreme heat slap him in the face. Startled, he dropped the bottle. It smashed on the floor, sending plutonium solution and shards of glass flying. It was 10:35 P.M. Instantly, the building criticality alarms sounded, and everyone upstairs dropped what they were doing to flee though the escape tunnel. The criticality was so extreme that it set off the criticality alarm in an adjacent building. The operator, knowing what had happened, proceeded to the decontamination facility.

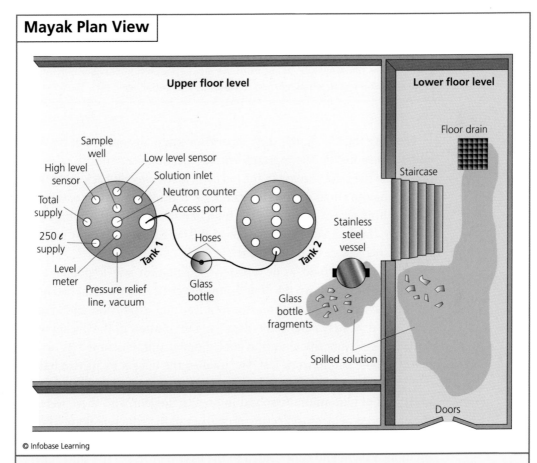

Mayak Plan View

© Infobase Learning

The floor plan of a criticality accident at the Mayak Production Association in 1968 is shown in this diagram. Workers transferred a solution of plutonium from tank 2 into the stainless steel bucket, and it became a runaway nuclear reactor.

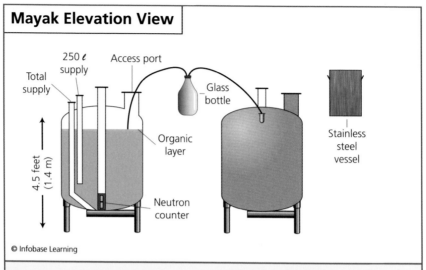

Mayak Elevation View

250 ℓ supply
Access port
Total supply
Glass bottle
Organic layer
4.5 feet (1.4 m)
Neutron counter
Stainless steel vessel

© Infobase Learning

This shows the Mayak accident from the side—the shift supervisor only made the situation worse when he tried to clean up the initial accident.

Hearing the alarm, the shift supervisor knew instantly that he would be held responsible for this unfortunate occurrence, and he wished to deactivate the critical mass in the bucket. He tried to get back in the room to see what he could do, but the radiation control supervisor was correct to deny him access. In spite of the high gamma-radiation readings at the doorway, the shift supervisor deceived the radiation supervisor into leaving the area and dashed into the room.

All he needed to do, as he assessed the situation, was to pour the contents of the bucket down the floor drain. The criticality incident had heated up the solvent and plutonium mixture, causing expansion and shutting down the delicately balanced criticality. The solution in the bucket would remain barely subcritical until it cooled down and reassumed its original density, so he had to work fast. He grabbed the bucket by its two handles and lowered it to the floor. Unfortunately, the floor was covered with a great deal of plutonium dissolved in organic solvent, and there were also the enhanced neutron-reflecting properties of the concrete floor. The plutonium on the floor not only reflected neutrons back into the bucket, it multiplied them and contributed to the fission process. The shift supervisor caught the blue flash right in the face as the mixture went supercritical. He never made it to the floor drain. Again, the criticality alarm went off in the next building.

Both personnel, the operator and the shift supervisor, were flown to Moscow for treatment of acute radiation sickness on December 11. The operator had received a 700-rem (7-Sv) dose. Both his legs and a hand were amputated, but he was still living 31 years after the incident. The shift supervisor absorbed 2,450 rems (24.5 Sv) of radiation, and he died one month later. The organic solvent research was discontinued.

CRITICALITY MISHAPS IN THE UNITED STATES

There seemed to have been disproportionately large numbers of criticality accidents at the Mayak plant in Russia compared to the United States, where the magnitude of plutonium and uranium production was about the same. Still, there were accidents in the United States, and one of the most interesting incidents occurred at the Y-12 facility at the Oak Ridge National Laboratory on June 16, 1958.

The Y-12 plant had been the center for electromagnetic isotope separation during World War II, but the buildings were reconfigured for chemical separation of enriched uranium from solid wastes after the war. Building 9212 was divided into two wings, B-1 and C-1. The entire facility, including Building 9212, had been shut down for an inventory of fissile materials, and only the process in the B-1 wing had been restarted. At 1:00 A.M. on Monday, June 16, the C-1 wing supervisor instructed an operator to drain the uranyl nitrate out of a glass standpipe at the pH adjustment station. Uranyl nitrate is a distinctively yellow fluid, and people who had worked in the fuel-processing industry were familiar with it. The C-1 wing normally received uranyl nitrate through a pipe from the B-1 wing, but the process in the C-1 wing was still shut down, so the gravity feed from B-1 was shut off with a valve, labeled V-1.

At 5:00 A.M., the supervisor noted that the glass standpipe was still full of uranyl nitrate, and he quizzed the operator as to why it had not been drained. The operator assured him that it had. With a trace-back of the pipe leading to the standpipe, he found that valve V-1 had been leaking. He gave the handle a hard turn and redrained the standpipe. At 7:00 A.M., the shift changed, and the outgoing supervisor did not mention that valve V-1 had been stuck open.

The purpose of the uranyl nitrate line coming from the B-1 wing was to fill three five-inch (127 mm) vessels. These vessels were purposefully designed to be too small to contain a critical mass of uranium, regardless of its level of enrichment or its concentration in a solution. They had been cleaned days before and now it was time to leak test them. The test was simple. An operator filled each vessel with water, and he looked for leaks. At 8:00 A.M., the supervisor assigned two operators the task of leak testing. They found no leaks, but now they had the three vessels full of water, and it had to be drained out. Rather than drain it out onto the floor, they rolled a 55-gallon (208-l) drum under valve V-11 on the uranyl nitrate line, connected a plastic hose to the valve, and proceeded to drain the contents of the three vessels into the drum.

What the operators did not realize was that valve V-1 had been leaking uranyl nitrate from the B-1 wing for several hours before it was tightened up, and one of the vessels was completely full of it. When they opened the drain valve, V-11, the heavy uranyl nitrate was the first fluid to drain, and it went straight into the drum. Shortly before 2:00 P.M., they opened the valve, and one of the operators stood by the drum to monitor the situation for any unusual conditions. Although he had been on the job for six years and certainly knew the difference between uranyl nitrate and water, he did not notice what was pouring into the drum. It took 15 minutes for the level of uranyl nitrate to reach criticality. The operator noticed curious yellow-brown fumes rising out of the drum. He took two steps back and suddenly saw the blue flash, characteristic of a reactor suddenly going critical. A quantity of 15 gallons (56 l) of enriched uranium-235 in nitric acid had accumulated in a geometry favorable to fission. The recording radiation meter in the building ran off the end of its scale, and everyone in the building left promptly.

The first blast of criticality heated the solution to the point where expansion shut it down, but it started again several times, in a pulsing action, in the next 2.6 minutes. More uranium was being added to the contents of the drum, along with water, as the valves were not closed before the operators fled the scene, and the fissions continued to pulse for the next 15 minutes. The drum finally went subcritical when the water from the leak testing had diluted the solution to the point of not being able to support a self-sustaining reaction. Eight people in the C-1 wing of Building 9212 received significant radiation doses, ranging from 28.8 to 428 rem (0.288 to 4.28 Sv). One survived 14.5 years, one 17.5 years, and five were alive 29 years after the incident. The status of one is unknown. The use of 55-gallon (208-l) drums was strictly curtailed at the Oak Ridge National Laboratory.

An even more serious accident occurred at the Los Alamos Scientific Laboratory in New Mexico on December 30, 1958, and the circumstances are odder than in most incidences. A 264-gallon (1,000-l) stainless steel vessel was being used to purify and concentrate plutonium from slag and crucible residues by dissolving it in aqueous and organic solutions. The vessel was of a size and shape favorable to criticality, with its rounded ends and soup can proportions, but this was realized and such vessels were scheduled to be replaced with slender vessels, six inches (15 cm) in diameter and 10 feet (3 m) long, in the next few months. Although the metal vessel was opaque and did not show the level of fluid inside, a sight-glass was connected directly to the side of the vessel, showing the height of liquid. A stirring propeller was in the bottom of the vessel, turned by an electric motor on top.

At 4:35 P.M., near the end of the workday on the last day before the New Year's holiday, there were two solutions in the vessel. An aqueous solution was in the bottom of the vessel, and it contained about

LASL Accident

Thermocouple well

Sight port

Sight port

Sight gauges

160 ℓ organic

330 ℓ aqueous

© Infobase Learning

At the Los Alamos Scientific Laboratory, a fatal accident occurred in 1958 when too much plutonium solution wound up inside this tank.

0.13 pounds (60 gm) of plutonium in 87 gallons (330 l) of water. An organic solution was floating atop the aqueous solution, containing 6.8 pounds (3.1 kg) of plutonium in 42 gallons (180 l) of solvent. The two fluids, organic and aqueous, would not mix unless forced. The aqueous solution of plutonium was diluted, and was in no danger of going critical. The organic solution, on the other hand, was very concentrated, but it was in an inefficient, shallow-cylinder configuration and far from critical.

There were three operators in the room. One bent over to look into the sight glass, close to the vessel, as he turned on the stirrer. The stirrer accelerated to top speed, 60 revolutions per minute, quickly. The aqueous solution in the bottom of the vessel with the stirrer spun rapidly, and the centrifugal force on the fluid caused it to change shape. The fluid rode up on the inside of the vessel, forming a spinning vortex. The organic fluid on top had to follow the aqueous solution as it changed shape, and it fell into the inverted cone of the vortex. This made the organic solution geometry more compact, and it was surrounded on the sides by aqueous solution. The aqueous solution made a fine neutron reflector, and the 6.8 pounds (3.1 kg) of plutonium was now in an advantageous shape. There was no blue flash visible, as the reaction was completely enclosed in a stainless steel vessel, but the operator could see violent boiling in the sight-glass.

The operator closest to the vessel received a dose estimated at 12,000 rems (120 Sv) and died 36 hours after the incident. Two other persons received doses of 134 and 53 rems (1.34 and 0.53 Sv), and they suffered no apparent ill effects. No radiation alarms in the building indicated a problem. Before the process was restarted, the program to replace the vessels was accelerated and installation was completed. There was nothing to clean up, no radioactive debris contaminated the workplace, and no equipment was damaged, and yet a man had died because of where he was standing when he threw a switch. The dangers of working in fuel processing could be subtle, and an unusual level of vigilance was necessary for safe operation of the plant.

A criticality at the Hanford Works near Richland, Washington, on Saturday morning, April 7, 1962, marked the first use of a robot as a response mechanism in a nuclear accident. An experimental plant had been built for testing the Recuplex process for extracting plutonium from used reactor fuel. A tank overflowed, putting an aqueous solution of plutonium on the floor of the room. An operator, against all instructions, opened a valve in the floor sump to clean up the solution and managed to put the solution back into the tank. This would seem a safe move, but there was already water on the floor when the plutonium solution collected there, and the additional moderating action of the water made the mixture supercritical back in the tank. Three people at the station received significant radiation doses of 110, 43, and 19 rems (0.11, 0.43, and 0.19 Sv), but the 19 other people in the building heeded the criticality alarm and evacuated quickly. The plutonium mixture in the tank remained supercritical for 37.5 hours.

The remotely controlled robot, equipped with a television camera, reconnoitered the building interior, fixed precisely the point of the accident using a directional gamma-ray counter, read meters, and operated the valves to shut down the operation. The use of the robot was impressive, and it would be used for other nuclear emergencies in the United States.

In the small village of Wood River Junction, Rhode Island, a privately owned nuclear fuel–processing plant was built. Operations began on March 16, 1964, and the plant was plagued by the expected number of start-up difficulties. In the extraction process, as a final scrubbing to remove uranium from the nitric acid solution, trichloroethane, or TCE, was bubbled through the column. To then remove the uranium from the TCE, it was transferred into five-inch (127 mm) polyethylene bottles, where it was mixed with sodium carbonate and manually shaken. On July 16, the operators had grown weary of this manual operation and asked to transfer this operation into the carbonate makeup tank, atop the solvent extraction column. The tank was 18 inches (457 mm) in diameter and 26 inches (670 mm) high. The supervisor gave his permission, and this new procedure was communicated to the next shift of operators.

On the day before the accident, on July 23, an evaporator column stopped working, and the shift operators dismantled it for cleaning. They found it stopped up by a plug of uranyl nitrate crystals. They dissolved it with steam, catching the resulting solution in a five-inch (127 mm) bottle, identical to the ones used for TCE. It was labeled as containing a concentrated uranium solution, to distinguish it from the other bottles.

On Friday, July 24, an operator needed an empty bottle and could find only full bottles. He asked the supervisor if he could empty a TCE bottle into the carbonate makeup tank. He was told no. Proceeding anyway, he picked up the bottle containing concentrated uranyl nitrate, not noticing the label, and climbed the stairway to the third level, over the solvent extraction column. The operator had almost completely emptied the bottle into the tank when the room was lit up by the blue flash. About 20 percent of the fluid in the tank erupted and hit the ceiling, the walls, and the operator, who fell off the stairs and hit the floor. He gained footing and ran to the decontamination building.

The criticality alarm was still blaring an hour and a half later when the plant supervisor and the superintendent entered the building to empty the makeup tank. The superintendent climbed to the top of the column and found the bottle, still upended in the tank. He removed the bottle, turned off the stirrer, and left the room. Just as he passed the supervisor, standing in the doorway, the tank achieved criticality a second time, dousing the supervisor with 100 rads (1.0 Gy) and the superintendent with 60 rads (0.6 Gy) of radiation. When he turned off the stirrer, the inverted conical vortex in the rapidly spinning fluid relaxed, and in a few seconds the shape of the uranium solution in the tank had become more compact and advantageous of criticality.

The operator had received 10,000 rads (100 Gy) of radiation, and he died 49 hours later. There was no physical damage to the facility, although an extensive cleanup of the spilled uranium solution was necessary.

MORE CRITICAL EXCURSIONS IN RUSSIA AND JAPAN

Novosibirsk is the administrative center of the Siberian Federal District of Russia, founded in 1893. It is also the site of the Siberian Chemical Combine, which is actually a large gaseous diffusion plant enriching uranium-235, 24 hours a day.

The gaseous diffusion method of uranium isotope separation involves uranium-hexafluoride gas, and it must be purified by freezing it to condensation temperature and then drawing off any air or hydrogen fluoride contamination using a vacuum pump. The vacuum pump was on a metal

Radiation is still being monitored in southwest Siberia, long after the fuel-processing plant at Novosibirsk was shut down. *(© Georges de Keerle/Sygma/Corbis)*

base six feet (1.9 m) long, bolted to the floor of the facility. The pump, which was driven by an electric motor through a rubber belt, was 3.5 feet (1.08 m) high, and above it was a gravity-fed, cylindrical lubricating oil tank, 2.5 feet (0.752 m) high. The oil in the tank had to be changed every 15 days, as it would gradually collect uranium-hexafluoride off the cylinder walls in the pump. The uranium-hexafluoride contamination would cause the oil to thicken and would degrade the efficiency of the pump.

The sedimentation cylinders feeding this vacuum pump were cooled by liquid nitrogen, but on July 1, 1961, an equipment failure shut down the liquid nitrogen production facility, greatly reducing the available supply. On July 2, the procedures were changed to accommodate the lack of automatically flowing liquid nitrogen, and operators were instructed to load the cylinders manually, using hand-carried Dewar flasks of liquid nitrogen. In violation of this directive, the cooling of the three intermediate cylinders was curtailed completely. This caused uranium-hexafluoride in the system not to be trapped in the cylinders, and it started feeding directly into the vacuum pump.

On July 10, an air-pressure regulator in the enrichment cascade failed, and this caused a greater than normal fraction of the uranium-hexafluoride stream to divert into the cylinders, which were not being cooled. The enrichment level of the uranium-235 was 22.6 percent. Natural, mined uranium is less than 1 percent uranium-235, and the high level of enrichment greatly improves the probability of fission and decreases the mass and volume required for criticality. At 4:45 A.M., the oil tank on the vacuum pump had accumulated enough uranium in the hydrogen-rich oil, a good neutron moderator, and it went critical, setting off the radiation alarm in the adjacent room and shutting down the process. The building criticality alarm did not sound.

The excursion was small, as the uranium in the oil was just barely critical, and the reactor shut down as the oil heated and the density of the mixture dropped accordingly. There was difficulty in finding the source of the radiation alarm. The radiation control officer surveyed everything in the room with the alarm but could not find any source of high radiation. He did not bother to go in the room with the vacuum pump, as it was never suspected of having a uranium buildup. At 7:30 A.M., he authorized a system restart. The operator started the vacuum pump and walked toward it to open a valve. Just then, he saw a blue flash, and this time the building criticality alarm sounded. Cold-starting the vacuum pump had

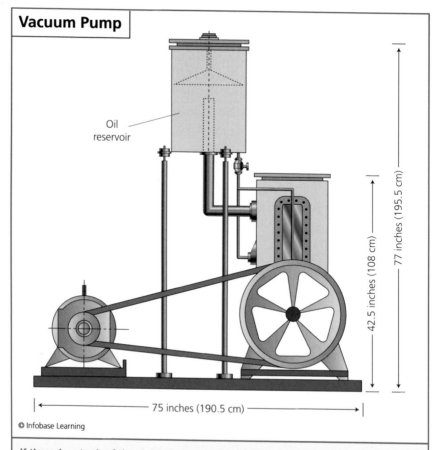

Vacuum Pump

Oil reservoir

42.5 inches (108 cm)

77 inches (195.5 cm)

75 inches (190.5 cm)

© Infobase Learning

If there is a tank of the right size and shape for a nuclear reactor somewhere in a fuel processing facility, then uranium will eventually find it. In this case, it was the oil reservoir on a vacuum pump at the Siberian Chemical Combine in 1961.

caused oil mixed with uranium-hexafluoride to collect in the standpipe in the middle of the oil tank, and a significant criticality resulted.

The operator absorbed a dose of 200 rads (0.2 Gy), and he suffered mild radiation sickness symptoms. Ever since July 1, 1961, multiple procedures had been violated, including failure to load liquid nitrogen into the cylinders and turning off the temperature recording instruments, and under these conditions accidents are likely.

An accident at the same plant on January 30, 1963, proved that criticality could occur because of the smallest mistake. In the metal production building, uranium was recovered from process wastes. The collection vessels for aqueous uranium waste solutions were large enough, 13 gallons (50 l), to support a criticality, and they were of optimum shape, with rounded ends. The only safeguard against collecting a critical mass in a vessel was frequent chemical analysis. Despite the importance of these analyses, there were two allowed reporting formats: as the percentage ratio of kilograms of uranium per kilogram of compound, or as the ratio of grams of uranium per kilogram of compound. The subtle difference in the two formats was a factor of 10.

On the day of the accident, uranium precipitate was dumped into collection vessel 64-A, and an analysis was requested. The results were 180 grams of uranium per kilogram of compound, but due to confusion over the format, the results were written down as 18 grams per kilogram. The operator added uranium compound to the tank, thinking that he now had 126 grams of it in 64-B, when actually he had 1,260 grams. Just to make sure, he asked for an analysis over the phone.

As luck would have it, the lab sent him back the analysis for a tank other than 64-A, and it happened to be exactly 10 times less than the actual concentration in 64-A. More compound was added, bringing the total amount of uranium to 5.5 pounds (2.5 kg) in the tank. The contents of 64-A were then transferred to an adjacent, larger tank, 64-B, and during the transfer the criticality alarm started to sound. The time was 6:10 P.M. Everyone evacuated the building.

Over the next 10 hours, the criticality cycled eight times, as the uranium compound heated under fission, expanded, stopped fissioning, cooled, and regained criticality. The fissions finally came to an end at 4:30 A.M. the next morning by an operator turning a valve and partially draining the tank. Work resumed by 6:00 A.M. Four people received radiation doses from six to 17 rads (0.06 to 0.17 Gy).

The last reported accident at Novosibirsk was on May 15, 1997. The plant had been producing enriched uranium fuel for the past 13 years without any trouble, but in 1996 during a routine inspection a solid uranium-oxide deposit of 12 pounds (5.5 kg) was found in a collection vessel. It was cleaned out using nitric acid, and nothing more was thought of the finding. It had probably taken all 13 years of operation for the deposit to slowly build up. All the vessels in the building, including the two receiver vessels, 59-A and 59-B taking a feed from the collection vessel, were specifically designed to discourage criticality in any concentration or volume of uranium up to 36 percent enriched.

However, there had not been any 36-percent uranium processed in Building 17 since 1986. The uranium being fabricated now was 90 percent uranium-235, and at 10:55 A.M. on May 15, the criticality alarm indicated that there was a solid buildup in vessels 59-A and 59-B. Everyone evacuated safely, and a recovery team went in and introduced a borated solution into the vessels to stop the criticality. Despite this action, the alarm went off again at 6:50 P.M. the same day, then at 10:05 P.M., and at 3:27 A.M. and 7:10 A.M. the next day, May 16. It was a most persistent critical mass. The

An aerial view of the JCO nuclear fuel processing plant in Tokaimura, Japan (AP Images)

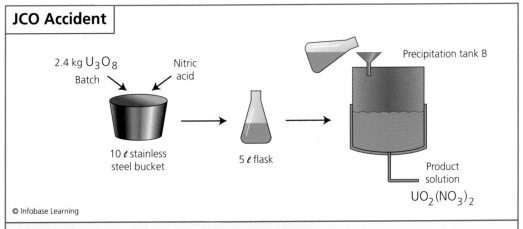

JCO Accident

2.4 kg U_3O_8
Batch

Nitric acid

Precipitation tank B

10 ℓ stainless steel bucket

5 ℓ flask

Product solution
$UO_2(NO_3)_2$

© Infobase Learning

Workers at the JCO plant put a uranium nitrate solution in the wrong vessel. Instead of putting it into many small bottles, they dumped it all in one large tank, with fatal results.

vessels were replaced by units built for 90 percent uranium, and a regular program for monitoring solids building up in tanks was instituted.

On September 30, 1999, the latest fuel processing accident occurred in the fuel conversion test building at the JCO Company in Tokaimura, Ibaraki Prefecture, Japan. It was the first and only fuel-processing criticality accident in which people outside the plant, on public ground, were subjected to measurable radiation exposure.

Three buildings at the JCO complex were licensed for processing uranium. Two main buildings were engaged in processing 5 percent enriched fuel. The fuel conversion test building was used only for a special project, processing 20 percent enriched fuel for the Joyo experimental breeder reactor at the Japan Nuclear Cycle Development Institute. On September 29, three operators began the task of preparing and packaging 20 percent enriched uranyl nitrate. The 37 pounds (16.8 kg) of uranium was to be shipped in one-gallon (4-l) bottles. A criticality accident was not considered a credible accident in this facility, so there were no criticality alarms, nor were the operators aware of the danger of transferring a uranium solution from a geometry unfavorable to criticality to a favorable geometry.

The method for draining uranyl nitrate into shipping bottles was spelled out in specific terms in the plant procedures manual. Uranium-oxide and nitric acid were to be dumped into a dissolver, which would drip the resulting uranyl nitrate into a long, thin vessel. Bottles were to be filled from a valve at the bottom of the vessel. Unfortunately, there was only a four-inch (10-cm) clearance under the valve, and it was hard to get the bottle under it. The operators improvised, mixing the chemicals in a

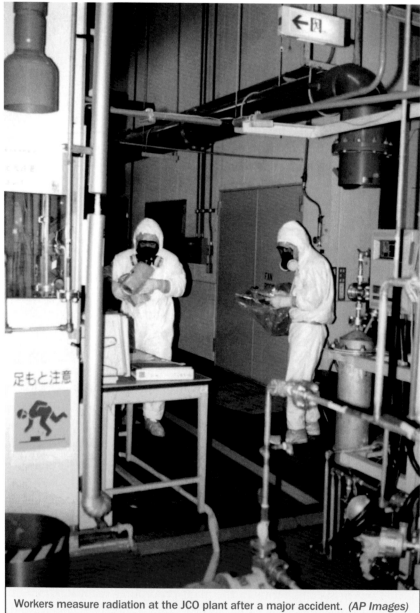

Workers measure radiation at the JCO plant after a major accident. *(AP Images)*

bucket, and then transferring the uranyl nitrate to a large precipitation tank using a 1.3-gallon (five-l) flask.

It took an entire day to process four batches of uranyl nitrate in the bucket. The next day, September 30, they prepared the final three batches. One operator sat at a desk nearby as the other two climbed the steps to the top of the precipitation tank. One held a funnel over an open port on the top while the other hoisted up the flask and tilted it to empty the contents. On the seventh batch, just before the last of the liquid was poured, at 10:35 A.M., the radiation alarms in all three buildings started sounding. Workers evacuated all buildings according to emergency plans, but they had to be moved off-site when the background radiation was measured. It was dangerous to be anywhere near the test building. All residents living within 1,148 feet (350 m) of the plant were evacuated, and residents within six miles (10 km) were advised to stay inside.

The power excursion in the precipitation tank continued for 24 hours, as officials tried to figure out what was happening. The JCO Company had no preparations for a criticality accident, and expertise and neutron detectors had to be imported from nearby nuclear power facilities. Experts decided to shut down the reaction by draining water out of the cooling jacket surrounding the precipitation tank, denying the critical mass the neutron reflections from the water. Opening the drain did not remove all the water, and the reaction continued. Finally, they routed pressurized argon gas into the pipes and that blew out the water and shut it down.

The operator holding the funnel received about 2,000 rads (20 Gy) of radiation, and he died 82 days later. The operator holding the flask received 1,000 rads (10 Gy) and died 210 days later. The operator sitting at the desk was exposed to 450 rads (4.5 Gy), and he was able to leave the hospital three months later. The 200 evacuated residents received less than 500 millirems (five mSv) of radiation. The JCO Company lost its license to process nuclear fuel.

The problem at the JCO plant was a lack of knowledge of nuclear fuel criticality accidents. If the managers, engineers, and workers had read this chapter, then this fatal accident would not have occurred. The next chapter discusses the chronology and causes of a significant nuclear accident in the United States, the Three Mile Island reactor meltdown in Pennsylvania. Repercussions of this unexpected event in 1979 still affect the civilian nuclear power industry, and it deserves detailed study.

7 Meltdown at Three Mile Island in Harrisburg, Pennsylvania

By 1970, nuclear technology seemed to be settling into a period of practical commercial power production, and most of the risky military experiments with nuclear weapons, aircraft engines, and compact or mobile power systems were either ended or toned down. Even the quest for a liquid metal–cooled fast breeder reactor had reached a resting point. In the 20 years since 1950, there had been a period of wide-open pursuit of all possible aspects of nuclear power, and in this unknown territory were discovered new and unique ways to have an accident. It took time, experience, and accident analysis, but eventually the new ways of reducing death and injury were implemented, and these measures became ingrained in nuclear technology.

The concept of civilian power plants using nuclear fission was completely different from any military application. Levels of risk to human health and property had to be reduced to as near zero as possible and simply locating a nuclear power plant in an uninhabited area was no longer a solution to the possibility of system failure. Nuclear systems would have to be located all over the country for power production, and proximity to industrial sites was advantageous. It reduced the waste of long-range electricity transmission and it reduced the expense of transmission facilities and right-of-way. Being close to human habitation, the fission reactors would have to be provided absolute radiation shielding, protecting everything outside the plant boundaries from dangerous exposure in the event of the worst possible system breakdown. To ensure the public acceptance of this new and unique power source, the plants had to be designed to be safer and cleaner than the existing coal and natural gas power plants.

The effort to achieve this level of public safety was evident in the relatively quiet period of the 1970s, when commercial nuclear power plants were being built all over the world. There were increasing numbers of power plants, but decreasing numbers of reported accidents. When the military experimentation stopped, the accidents did also. So it seemed until near the end of the decade in 1979, when a quietly operating nuclear power plant built on an island in the Susquehanna River, the Three Mile Island Nuclear Generating Station, experienced a problem with its primary cooling system. The result was not the worst nuclear accident in a technical sense. There were no casualties, no massive radiation doses, and no fallout blanketing the ground, but in terms of public relations, it was the worst accident in the history of nuclear power in the United States. Unlike a military operation, training exercise, or experiment, there was no way to cover up this accident with secrecy or a cloak of national security. It was out in the open, immediately and completely, and everyone saw, for the first time, that an unfortunate series of mechanical failures and operator missteps could destroy a multi-billion-dollar public investment.

This chapter covers the Three Mile Island accident with a detailed, step-by-step analysis, preceded by an explanation of the technology involved. The effects of this accident were largely psychological and the sidebar goes into the psychology and its ramifications.

THE WATER-COOLED REACTORS

The 1970s were a quiet period for nuclear power reactor accidents in the United States, except for one odd nonnuclear accident in the middle of the decade. On March 22, 1975, an accident happened at the Tennessee Valley Authority's *Browns Ferry Nuclear Power Station* near Decatur, Alabama, that would take the plant out of operation for an entire year.

Browns Ferry Unit 1, a General Electric boiling water reactor, was brought to power and connected to the grid on December 20, 1973. The reactor took cooling water from the Tennessee River on the north side of Wheeler Lake, and it was designed to deliver 1,065 megawatts of electricity.

In 1975, the reactor was down for upgrades, and a new set of cables had to run into the control room. Control room cables are always run up through the floor in a nuclear plant, and the room one story below the control room is the cable-spreading room. It is a room dedicated to nothing but the orderly routing of cables to the instruments, switches, warning lights, and alarms in the room above. A slight positive air pressure is kept in the room, so that dust will not collect in it, and all the penetrations for cables coming in from all over the plant must be sealed using polyurethane foam.

On March 22, new cables had been installed into a spreading room penetration. The seal had been broken to rout the new cables and then restored using a temporary compound, applied with an aerosol can. The worker needed to check the temporary seal for air leaks, so he lit a candle and observed the flame, looking for small air currents to flicker the flame. Immediately, the temporary foam caught fire, and with a great deal of effort the workers managed to put it out. The fire, however, had spread to the other side of the wall, inside the spreader room, and control room cabling for Units 1 and 2 was consumed in the blaze. The entire control room had to be re-cabled, and the job required a year of difficult labor.

An extensive investigation by the Nuclear Regulatory Commission (NRC) resulted in the publication of a new federal regulation, 10 CFR 50.48. This event proved pivotal not just for fire safety in the nuclear energy field but in all commercial and industrial construction. New fire-stopping sealants were developed and implemented, using silicon foam, and now a wide array of fireproof sealants are available for nonnuclear construction.

All three Browns Ferry reactor units were shut down by the NRC in 1985 for operational and management issues, and the power plant remained idle for six years. The cooling towers at Browns Ferry were not the tall, pinch-waist structures with which many are familiar, but were long lines of smaller cooling units, made of redwood slats, fiberglass, and aluminum. Water in this type of cooler is cascaded over the redwood louvers, producing a cooling effect for the reactor turbine exhaust. After sitting idle for a year, all the redwood had dried out in the Alabama sun, and a spark set one of the cooling units ablaze. The entire cooling unit array, as long as a football field, burned to the ground. In no case of fire at Browns Ferry was anyone injured, no radiation escaped, and no reactor fuel or machinery was damaged, but it was difficult to generate power at Browns Ferry.

In 1968, the General Public Utilities Corporation began construction of a pressurized water reactor power station in Londonderry Township, Pennsylvania, near Harrisburg, on Three Mile Island. Specifications called for two reactor/turbogenerators, having separate reactor containment buildings and cooling towers but a common turbine deck and control room area. The contractor was the Babcock & Wilcox Company (B&W) in Lynchburg, Virginia, an established builder of steam boilers and, as of 1961, pressurized water reactors. B&W had built components for the USS *Nautilus,* the world's first nuclear-powered submarine, in 1953, and in 1961 they built the entire reactor system for the first commercial nuclear ship, the NS *Savannah.* B&W built the Indian Point pressurized water reactor on the Hudson River in New York in 1962. The Three Mile Island Nuclear Generating Station was the second, and the last, civilian nuclear plant that B&W built.

The Three Mile Island Nuclear Generating Station near Harrisburg, Pennsylvania *(Bill Pierce/Time & Life Pictures/Getty Images)*

Commercial operation of Three Mile Island Unit 1 (*TMI-1*) began on September 2, 1974. The operating staff had been trained primarily in Admiral Hyman Rickover's nuclear navy program, with its rigorous and successful set of concepts on how a pressurized water reactor should be operated. A point that was stressed in navy training was the importance of maintaining the water level in the pressurizer. In addition to ensuring high pressure in the reactor vessel, the pressurizer also offered a large air bubble in the primary cooling system. The pressurizer was always to be

THE CHINA SYNDROME

In October 1966, the Fermi 1 fast breeder reactor in Michigan experienced a core meltdown when a piece of zirconium sheet metal broke free of its mounting and clogged coolant channels in the core. The nature of this accident caused a great deal of concern among nuclear engineers, as it was a core melt not in a water-moderated reactor but in a fast-neutron breeder. Any loss of coolant in a water reactor will automatically shut the reactor down, but this is not necessarily true in a breeder reactor. A loss of coolant and the resulting fuel melt in a breeder can actually improve the reactivity of the core, instead of degrading it, and the core can continue critical power production in the damaged condition. An engineer noted that the core in Fermi 1 could have continued to generate full power as it melted through the bottom of the containment vessel and kept going as it melted through the ground, "clear down to China" on the other side of the Earth. While his scenario was not really possible, his point was made. In the nuclear engineering community, this phenomenon was labeled "the China syndrome."

On March 16, 1979, Columbia Pictures released *The China Syndrome*, a film starring Jane Fonda, Jack Lemmon, and Michael Douglas, which successfully played to large audiences nationwide. In the film, Fonda is a television reporter and Douglas her camera operator, and they have both been assigned to film a story about a nuclear power plant near Los Angeles, California. The public relations director of the plant gladly shows them around, and they get to film the operation of the reactor from a glass-enclosed observation deck over the control room.

While they are filming, the plant experiences an emergency shutdown, or a scram, and the camera operator manages to get it all on tape. It is a serious incident, in which the water level in the reactor vessel seems to be rising too high, and the kept half full, so that if a shock wave were to travel through the system, it could be absorbed safely by the compressible air bubble. The sudden turn-on of a pump or the opening of a valve would cause a shock to the plumbing system, known as the dreaded "water hammer," and a shock of sufficient force could tear out sections of pipe or wreck valve mechanisms. It was drilled into the navy trainees, keep careful watch on the condition of the pressurizer, and the pressurizer will take care of the reactor. They were told to never let the pressurizer "go solid," having no air bubble at the top.

operators start opening valves to bring the level down. Before long, they realize that the needle of the water level indicator is stuck, and the level had been normal before they started dumping water. They were fortunate to catch the error before the top of the core was uncovered with coolant and the fuel started to melt. A full investigation by the NRC results.

The shift supervisor of operations is played by Jack Lemmon, and he is troubled by this incident and other related mechanical problems at the plant, as the television reporters grill him off-site about the emergency they had witnessed from the observation deck. He begins his own investigation and finds to his dismay that X-rays of the pipe welds in the primary cooling system, supposedly taken to ensure the integrity of welded pipe joints, had been fraudulently filed to save construction money. Becoming ever convinced of the lack of safety in this nuclear power plant, he steals a pistol from a security guard and takes the entire reactor control room hostage, forcing the operations staff to shut down the reactor.

In this film, the roles of good and evil are sharply drawn. The power company that owns the reactor is in business only to make money with safety issues largely ignored, and the television reporters are crusaders for public safety. In addition to the China syndrome concept, the film also borrows from the Fermi 1 engineering analysis the observation that if the containment vessel were punctured, a radioactive cloud could cover an "area the size of Pennsylvania." Although in the movie, people are harmed only by the power company's professional hit squad, using high-power rifles and an unmarked van, it generates a measurable concern about commercial nuclear power plants. It was a very popular movie, nominated for four academy awards, and it caused everyone who watched it to have second thoughts about the move toward an ongoing conversion of the nation's power source from coal to nuclear.

Twelve days after the movie opened, the incident at Three Mile Island occurred.

A FAILED PUMP, A STUCK VALVE, AND A MELTED CORE

The Three Mile Island Unit 2 (TMI-2) reactor first went online on March 28, 1978, at 4:00 A.M. Exactly one year later, Three Mile Island Unit 1 (TMI-1) was down for refueling, and at 4:00 A.M. TMI-2 was operating at full power. Suddenly, the main feed-water pump in the secondary cooling loop stopped pumping. Although it has never been established exactly why the pump stopped working, the reason is not important, because a nuclear plant is configured to automatically handle such a problem on a routine basis. The failure may have been due to water entering a pneumatic line controlling the flow through the condensate polisher filter.

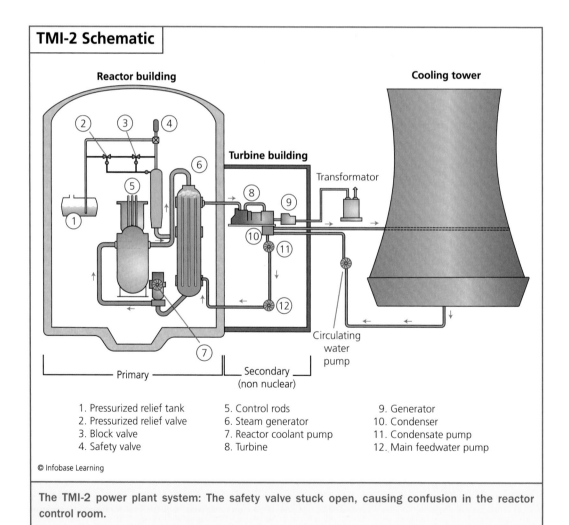

TMI-2 Schematic

Reactor building

Cooling tower

Turbine building

Transformator

Primary — Secondary (non nuclear)

Circulating water pump

1. Pressurized relief tank
2. Pressurized relief valve
3. Block valve
4. Safety valve
5. Control rods
6. Steam generator
7. Reactor coolant pump
8. Turbine
9. Generator
10. Condenser
11. Condensate pump
12. Main feedwater pump

© Infobase Learning

The TMI-2 power plant system: The safety valve stuck open, causing confusion in the reactor control room.

Immediately, the steam turbine automatically shut down, and then the electrical generator kicked itself off the power line. The reactor scrammed normally and went to standby mode. By this time, the operating staff looked up at the light panels above the control console to get a reading of the situation, and they began canceling alarm annunciators. The primary cooling loop, no longer having a flow through the secondary loop to remove the heat, over-pressurized, and the pressure relief valve blew open, venting the excess contents of the pressurizer into the relief tank.

To this point, all emergency operations were normal, and it has been said many times that if the operating staff had somehow accidentally locked themselves out of the control room, the Three Mile Island accident never would have happened. The pressure was relieved immediately, and a light came on at the control console indicating that a signal had automatically been sent to reclose the valve. The valve stuck open, regardless of the signal. There was no light to indicate that the valve was, in fact, still open, but since the signal had been sent, it was just assumed that the valve had closed.

Pressure continued to fall in the primary cooling loop, as the relief tank started to fill with coolant, and the water boiled furiously in the pressurizer. The operators carefully monitored the water level in the pressurizer, but the boiling action confused the level sensor, giving an optimistic reading. Meanwhile, three emergency feed-water pumps had started automatically in the secondary loop, but for some wrong reason two valves were closed in the emergency feed-water lines, and coolant could not reach the steam generator. It took eight minutes to sort this out, and by the time they had the emergency feed-water flowing correctly, steam bubbles had formed in the primary loop. Heat was not transferring from the core to the steam generator and out the containment vessel through the secondary loop.

A big steam bubble formed in the reactor vessel, because the pressurizer valve was still wide open, and this pushed water up into the pressurizer. The operators saw the water level rising in the pressurizer, and to prevent it from going solid, they turned off the emergency core cooling pumps, which had come on automatically when the pressure dropped.

At 4:11 A.M., the relief tank had reached full capacity, and its alarm sounded. The temperature indicator on the pipe from the relief valve to the relief tank showed an unusually high reading. The pressure and the temperature in the containment vessel were both increasing abnormally, and

TMI-2 Containment Building

Containment building

Reactor vessel

Fuel rods

© Infobase Learning

A cutaway drawing of the TMI-2 concrete containment building. This building prevented any uncontrolled escape of radiation from the plant during the accident.

this seemed to indicate a loss-of-coolant accident, but these indications were ignored by the operators, who were busy dealing with the water level in the pressurizer. At 4:15 A.M., the emergency diaphragm on the relief tank blew out, and it began dumping water out onto the floor, draining into the containment sump. An annunciator in the control room announced the presence of water in the sump, adding one more to the collection of alarms. Automatically, the sump pumps came on and began draining the excess water into an auxiliary building, but the pumps were stopped manually at 4:39 P.M.

After almost 80 minutes of slow temperature rise, the primary coolant pumps began sucking air, as most of the water in the system was now in the containment sump. The operators turned off the pumps, believing that natural thermo-siphon circulation would keep the coolant flowing. The top of the fuel was now exposed, with no coolant touching it, and the hot zirconium fuel rods started to corrode. Fission products began dumping into the remaining coolant, and zirconium was sucking oxygen out of the water, leaving hydrogen gas. The hydrogen leaked out through the broken diaphragm on the relief tank, mixed with the air in the building, and later that afternoon, it exploded. A commercial reactor containment building is sufficiently robust to sustain an internal hydrogen explosion without damage. The muffled sound and the slight vibration in the control room floor were hardly noticed.

At 6:00 A.M., the operations shift changed, and the fresh operators were apprised of the situation. One of the new arrivals noticed that the relief tank temperature looked extremely high, and he shut off the flow from the relief valve using a backup system. It was too late. Already 250,000 gallons (950 m³) of coolant had leaked onto the floor. At 165 minutes into the incident, the radiation alarms started going off as the water, now heavily contaminated with fission products, reached the radiation detectors.

At 7:00 A.M., the plant management declared a site area emergency, and at 7:24 A.M., it was upgraded to a general emergency, which was the highest level of emergency in the list of accident indicators from the NRC. Radio station WKBO, Harrisburg, announced a problem at Three Mile Island at 8:25 A.M., and the Associated Press picked it up at 9:00 A.M. By noon, the entire world knew about the emergency shutdown at TMI.

President Jimmy Carter in the Three Mile Island Unit 2 control room, assessing the accident *(AP Images)*

In the control room, it was still not clear what was happening in the reactor. There was no way to measure the coolant level in the reactor vessel. It had always been assumed that, because the pressurizer is mounted above the reactor, if the pressurizer has water in it, then the reactor is filled to

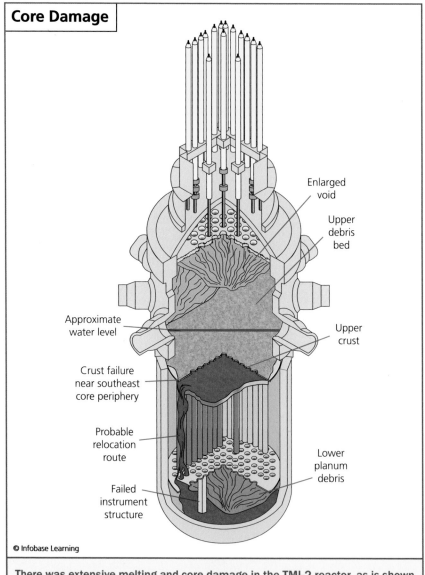

Core Damage

Enlarged void

Upper debris bed

Approximate water level

Upper crust

Crust failure near southeast core periphery

Probable relocation route

Lower planum debris

Failed instrument structure

© Infobase Learning

There was extensive melting and core damage in the TMI-2 reactor, as is shown in this diagram. No melted fuel breached the steel reactor vessel.

the top. A group of operators took readings from thermocouples at a variety of locations and obtained a sample of water from the primary cooling loop. It showed severe radioactivity, and the temperature readings were not normal. Finally, seven hours into the emergency, new water was pumped into the primary loop and the backup valve on the pressurizer was opened to relieve the vapor lock. After 16 hours, the primary cooling pumps were turned back on, and the core temperature started to fall. Almost the entire reactor core had melted. No one, in the power plant operating staff or the population of Pennsylvania, had been exposed to any radiation, but TMI-2 would never again generate electrical power, and it would take decades to repair something that was more valuable than a power plant, the damage to the public trust of the nuclear industry.

8 The Disaster at Chernobyl, Russia

The Union of Soviet Socialist Republics (USSR) showed an early interest in nuclear technology. Because of espionage from the Manhattan Project in the United States, the Soviets were able to mount efficient nuclear weapons and energy immediately after World War II, building test reactors and plutonium-production facilities. On August 29, 1949, they tested a plutonium-fueled atomic bomb, *First Lightning*, at Semipalatinsk, Kazakhstan. From that point, it seemed a constant race or competition between the Soviet Union and the United States, or the East and the West, for technical dominance in this and other areas of technical progress. Much of the breakneck speed of nuclear technology development was due to this rivalry.

The Soviets were able to claim the first civilian power reactor, the Атом Мирный, or Atom Mirny, meaning "Peaceful Atom," in *Obninsk*, Russia. Construction began on January 1, 1951, and the first start-up was on June 1, 1954, switching five megawatts of electricity into the local power grid. The power plant operated flawlessly, producing power until April 29, 2002, when it was finally shut down. It was a power plant reactor design never used for electrical generation in the United States, using a combination of a graphite moderator and a water coolant, but its success encouraged engineers in the Soviet Union to upscale the design, from a meager five megawatts to an impressive 1,000 megawatts. The resulting design was the *reaktor bolshoy moschnosti kanalniy (RBMK)*, the pride of the Soviet nuclear industry, capable of a dual purpose. It not only generated clean power for the growing Soviet industrial plant, it also produced plutonium for the Soviet military weapons stockpile. In the United States, the two processes of electrical power generation and plutonium production were considered to be different enterprises, one civilian and one military, and they were kept completely separate, with different safety standards, security concerns, and end products. The Soviets got all nuclear advantage out of one type of reactor.

An RBMK blew up at the Chernobyl Nuclear Power Plant on April 26, 1986, and by any measure it was the worst nuclear power disaster that has ever occurred. There are many lessons to be learned from this catastrophe. Perhaps one of the most fundamental was the potential for engineering folly when taking a successful design for a small machine, five megawatts, and making it larger, 1,000 megawatts. Power can be scaled up, but with it the risk and the danger also scale up. This chapter gives a clear description of the uniquely dangerous RBMK commonly used in Russia, the conditions leading up to the accident, a description of the explosion, and a discussion of the far-ranging aftermath of the Chernobyl catastrophe. A sidebar describes the evacuated town of *Prypiat,* where the power plant workers and their families once lived.

THE RBMK-1000 GRAPHITE REACTORS

RBMK is an acronym made from the first letters of the Russian description of a power reactor type, Реакор Болъшой Мощности Каналъныий, pronounced reaktor bolshoy moschnosti kanalniy, meaning "reactor (of) high power (of the) channel (type)." In Western terms, we would call it a graphite-moderated boiling water reactor. The bulk of the reactor, which is the size of a gymnasium, is graphite blocks, and its greatest advantage is that it requires no expensive enriched uranium fuel. It can burn natural uranium, refined straight from uranium ore mined out of the ground. Although neutron-moderating properties of the graphite are such that a critical mass can be assembled using this lowest-grade fuel, in 1986 the Chernobyl-4 reactor was running on low-enriched fuel, at 2 percent uranium-235.

Ordinary water, which is used in enriched-fuel reactors as both a moderator and as a coolant, is used as the coolant in an RBMK. The dominant volume of the reactor is graphite, so the detrimental properties of the water are insufficient to shut the reactor down, and no very

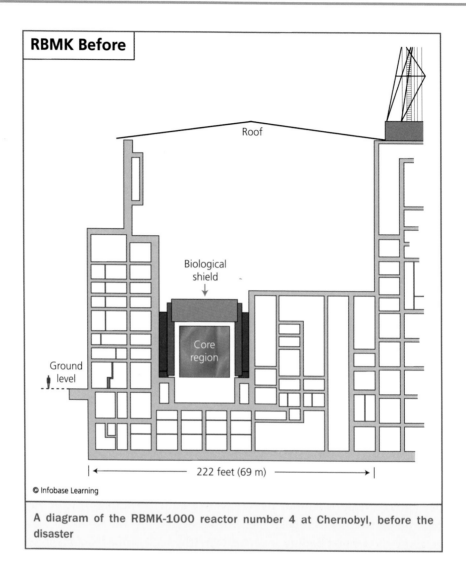

RBMK Before

Roof

Biological
shield
↓

Core
region

Ground
level

|← ——————— 222 feet (69 m) ———————→|

© Infobase Learning

A diagram of the RBMK-1000 reactor number 4 at Chernobyl, before the
disaster

expensive heavy water, or deuterium oxide, is necessary as a coolant.
An RBMK runs hot, at 554°F (290°C). At that temperature, graphite
annealing, the process that made operations so difficult for the British at Windscale, is not necessary. Control rods are inserted vertically,
driven in and out by electric motors on top of the reactor. Because the
RBMK is used for plutonium production, the fuel must be changed out
frequently, as overexposure to neutrons tends to contaminate the fissile plutonium-239 with the undesirable plutonium-240 isotope, and a
tracking crane is mounted atop the reactor for this purpose. The crane

takes up a lot of vertical space, and the resulting building is far too big for a Western-style containment vessel.

The sealed, rigid structure surrounding all civilian power reactors in the United States is designed to prevent leakage of any radioactive material, gaseous, liquid, or solid, from the radioactive processes in the power plant. These structures are built to withstand explosions of the reactor within the building and to withstand an airliner crashing into the building from outside. A Soviet RBMK is built with a roof over it to keep rain off the equipment and to prevent overflying spy satellites from seeing the frequency of the plutonium extraction.

Cooling water in an RBMK is run vertically through 23-foot- (7-m) long metal tubes in holes bored in the solid graphite. Cold water, pumped in through the bottom, heats as it rises in the tubes, boils, and is diverted to two steam separators on the sides of the reactor, behind radiation shielding. The steam is then sent to an adjacent building to run the turbogenerators, and it returns from the steam condensers for reintroduction to the bottom of the reactor. Also running vertically in the graphite are 3,000 fuel assemblies, consisting of metal tubes containing pellets of uranium dioxide. A metal plate at the bottom of the reactor keeps the fuel from falling out of the core, and a metal plate on top is perforated so that the fuel can easily be withdrawn for replenishment and plutonium extraction. An RBMK can be refueled while it is running and producing power.

There were economic advantages to building RBMK reactors, but there were also two safety problems with the design that were recognized even by the Soviet nuclear establishment. The first problem was the use of graphite. Graphite had been used successfully in the United States for plutonium production reactors, but the designs were based on preliminary information available in 1942. The British had also built graphite-moderated production reactors, but they had experienced a disastrous fire in Windscale-1. There was nothing to prevent such a fire in an RBMK, with the main difference being that an RBMK was a larger core with more graphite. More graphite simply means that a bigger fire is possible. The second problem was the well known "positive void coefficient."

In a water-moderated and cooled reactor, when the water boils, leaks away, or is blocked from running through the reactor, it creates a void in which there is no water. This condition, however it is started, will inevitably cause the reactor to shut down, or self-scram. Without all of the moderator, a reactor cannot maintain fission. Coolant and moderator are the same water, so the coolant is gone also, and this may or may not cause the fuel to melt, but at least it does not melt under power-production conditions, as the reactor is shut down. Because the power goes down when there is a water void, this effect is called the "negative void coefficient." A water void can also develop in a graphite reactor, but the moderator does not go away, only the coolant is lost. Regardless of how hot the reactor becomes with its coolant void, the moderator remains in place. Without the water, the moderating properties of the reactor are actually improved. Graphite is a much better moderator than water, so the reactivity of the core increases instead of decreases. Instead of going subcritical and shutting down, the graphite reactor goes supercritical, and the power increases, out of control. This is called the "positive void coefficient," and it is considered undesirable. If a reactor has lost its coolant, it is best to shut down instead of increasing power. At low power levels, the coolant is an unstable mixture of water, boiling water, and steam, with the water content adding a negative reactivity. If the core temperature rises, the water completely flashes to steam, and the moderation of the core improves, as water density in steam is very low. As the temperature or power level rises in an RBMK, the fission level can take off in a supercritical excursion.

These were the obvious problems. The most significant flaw in the RBMK design was the control rods, intended to absorb neutrons when inserted into the core and bring down the fission process. The control rods were made with graphite tips, and the ends of the rods were hollow and filled with water. This mechanical extension at the ends of the rods was three feet (one m) long, and the remaining length of each rod was made of boron carbide, a neutron absorber. The bored-out channel for each vertically inserted control rod fills with water as it is pulled out, and the graphite tip was supposed to counteract the negative reactivity of the water with the positive activity of graphite. Unfortunately, after a rod is fully withdrawn, for the first few seconds of rapid insertion the power level of the reactor does not go down, as one would expect. The power level goes up, because graphite is being introduced, and not boron carbide. Only when the boron carbide section of the rod slides into the top of the core does the fission rate start to go down. This operational quirk was not obvious, and it was unknown to the reactor operators.

The earliest design was the RBMK-1000, built to produce 1,000 megawatts of power, and reactors of this type were built in Chernobyl, Kursk, Leningrad, and Smolensk, typically with four reactors per site. Construction at Chernobyl, located near the town of Prypiat, Ukraine, about 68 miles (110 km) north of Kiev, was started in 1970. Chernobyl-1 came online in 1977, followed by Chernobyl-2 in 1978, Chernobyl-3 in 1981, and Chernobyl-4 in 1983. Two more reactors, Chernobyl-5 and -6, were under construction in 1986. The design was upgraded to the RBMK-1500, a reactor capable of generating a very impressive 1,500 megawatts of power, and plans were drawn for installations at Ignalina and Kostroma.

THE EARTHQUAKE IN JAPAN

Japan is located on the western rim of the "Ring of Fire," an area rife with earthquakes and volcanoes that encircles the Pacific Basin. It is on the leading edge of the tectonic plate of Asia. Meeting it from the east is the Pacific Plate, which is moving to the west at about the same speed as a thumbnail grows, shrinking the Pacific Ocean and moving North America ever closer to Japan. Where they meet, the Pacific Plate folds below the Asian Plate, and the tectonic movement is neither smooth nor continuous. When the plates crack loose, Japan is hit—about every 30 years—with a major offshore temblor.

This geological quirk makes Japan one of the last places on Earth one would choose to build a nuclear power plant, particularly on a beach. However, Japan has a vibrant, successful industrial economy that requires a lot of electrical power, and almost 30 percent of that power is generated through nuclear means. There are 54 nuclear power reactors in Japan, grouped into 17 power plants, and every one of them is located on the coast, directly vulnerable to tsunamis.

Construction of the Fukushima I Nuclear Power Station on the northeast coast of Japan began in 1967. When completed in 1979, the plant was capable of generating 4,696 megawatts of electrical power, using six boiling-water reactors designed by General Electric. Being keenly aware of the danger posed by earthquakes, Japanese engineers built the plant with extra bracing and strength. As with all Generation II power reactors, these were equipped with a second steel barrier, called the containment structure, to limit radiation release from the reactor core. If the stainless steel reactor vessel were broken, all the released steam and fission products would be stopped by this second vessel. Beyond the containment structure, the reactor building was the last defense between the highly radioactive reactor core and the world outside.

At 2:46 P.M. on March 11, 2011, reactors 1, 2, and 3, the oldest reactors at the Fukushima plant, were up and running. Reactors 4, 5, and 6 were down for maintenance

(continues)

(continued)

and refueling, and 6,415 people were on-site and working. Makoto Nagai, an inspector for the Nuclear Industrial and Safety Agency, noticed his coffee mug bouncing sideways across his desk; a magnitude 9.0 earthquake, the largest ever recorded in Japan, had erupted underwater, 80 miles (130 km) offshore. Within seconds, the solid, heavily braced power plant building was shaking violently. Fearing a falling ceiling, Nagai ducked under his desk, as bookshelf brackets became dislodged from the walls in his office.

All three reactors shut down automatically. In three seconds, all the control rods in the reactors were fully inserted, and shut-off valves in the steam lines to the generator buildings had slammed shut. The plant stopped generating electrical power. Six miles (10 km) west of the plant, the substation crumbled, cutting the power plant off from power generated anywhere else. Emergency lights came on.

Each of the six reactors was equipped with two emergency diesel-powered generators. There was one spare generator on-site, a total of 13 generators. All started automatically when the reactors shut down, providing on-site power to keep the coolant pumps running and the control rooms functioning.

Workers scrambled to dig out of tossed office furniture and assess the damage, but a little more than an hour after the earthquake shocks, a low rumble started, and the walls of the power plant began to vibrate. At 3:41 P.M. all the power ceased again. A wall of water 49 feet (15 m) high washed over the seawall and inundated the power plant. The generators, located in the basements of the turbine buildings, were flooded with seawater. Other nuclear plants were on the east cost, and some were even closer to the epicenter of the earthquake, but Fukushima I had the misfortune of being the plant that lost the use of its diesel generators.

Without electrical power, the emergency core cooling systems could not function. Remote-control valves could not be opened or closed, and the system monitoring instruments in the control room did not operate. Even if power could be restored, electrical wiring, junction boxes, switches, and pump motors were soaked in salt water. Coolant pumps stopped working.

Deprived of circulating water, units 1, 2, and 3 began overheating. Although the reactors were completely shut down, the delayed effects of fission were still producing heat. It was a small percentage of the heat produced when the reactors were making power, but active cooling was still needed. The water trapped in the reactor vessels boiled to steam, like a kettle left on a stove burner. Steam relief valves opened on the vessels to prevent steam explosions, and the steam ran into the Mark I containment structures. With no inlet water to replace it as it steamed away, the coolant in the three units began dropping.

By the next morning, Unit 1 had lost enough water at the top of the core to expose the fuel to air. With no coolant, the zirconium alloy fuel cladding became red hot, and it began to react chemically with the steam, producing zirconium oxide and hydrogen gas. With the cladding burned away, highly radioactive fission products in the fuel began to dissolve in the steam. Hydrogen and nuclear waste left the reactor vessel, along with the steam, into the containment structure. A few hours later, the dangerously high steam pressure in the containment was relieved by a valve opened manually. Hydrogen from the decomposing

The Unit 3 reactor building on March 24, 2011. The top of the reactor building has been demolished by a hydrogen explosion, and workers have hacked holes in the roof of the turbine building, shown in the top section of the picture, to prevent further hydrogen accumulation. What appear to be sticks in the remains of the frame of the building and scattered about are actually heavy steel girders. *(AP Images)*

fuel cladding escaped into the air in the reactor building. At 3:36 P.M. on Saturday, it exploded, sending the top of the reactor building flying. The roof disappeared, leaving the heavily damaged reactor and its spent fuel pool exposed to the open sky.

Over the next weeks, the situation only worsened, with hydrogen explosions, melted reactor cores, and fission products leaking unchecked into the environment of Japan, a country that was beset with the problems of recovering from an earthquake that killed 26,000 people. It was the worst nuclear accident in the 21st century, so far.

THE FATAL EXPERIMENT

Unit 4 was the newest operational reactor at Chernobyl, and it had been generating power for about three years without having run a required safety test. It was scheduled to be shut down for maintenance on April 26, 1986, and it was an opportune time to get this nagging omission on the books and out of the way. The day before, April 25, the procedure was set up, to test the ability of the turbogenerator to keep spinning and generate enough power to run the water pumps in the event of a sudden scram and a loss of steam from the reactor. The safety rule was that the pumps had to keep running under all emergency conditions, to keep the coolant flowing and prevent fuel melting, even if the reactor were in shutdown mode and could not give power to the turbines. It had to be assumed that under the worst condition the power grid would be down, and no power would be available from outside the plant. The plant had diesel generators poised to be started of sufficient power to run the pumps, but it took 40 seconds to get them started. The question was, would the turbogenerators have enough momentum built up to keep spinning long enough to give the diesels time to get started, if the only thing connected to them were the pumps?

The test had already been run before on one of the Chernobyl units, with negative results. The turbine lost speed quickly, and there was no way for it to power the pumps long enough to wait for the diesel backups. The turbines had been modified, perhaps fixing the problem, and now it was Unit 4's turn to try it. The most experienced operating crew ran the reactor during the day shift, and they were prepared to do the experiment. They gradually reduced the power of the reactor to the specified level for the experiment, but as they powered down another power plant in the Kiev region dropped off-line. The Kiev grid controller called and requested that they keep running at their current power level, 50 percent, to help with the evening peak demand for electricity, at least until the other plant came back online. The operators agreed. The test would have to be delayed until the next shift. While an RBMK was rated at 1,000 megawatts of electrical generation, it actually produced 3,200 megawatts of heat, with more than two-thirds of the power wasted into the cooling towers. At 50 percent, they were generating 1,600 megawatts of pure heat.

It was important to slowly reduce the power. A rapid shift in power level would allow xenon-135 to build up in the core. Xenon-135, a constantly produced product of fission, was a strong neutron absorber, and its presence tended to shut a reactor down. Criticality was maintained

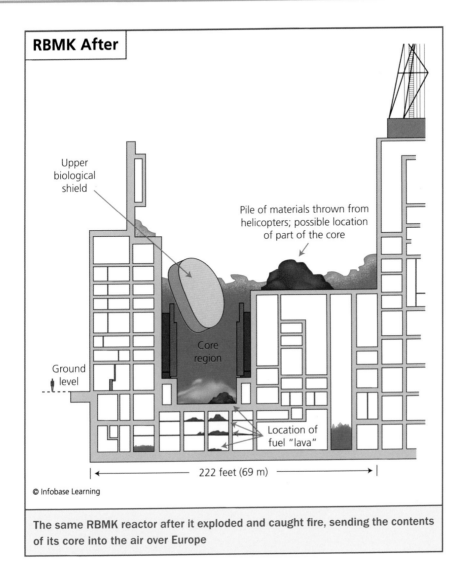

RBMK After

Upper
biological
shield

Pile of materials thrown from
helicopters; possible location
of part of the core

Core
region

Ground
level

Location of
fuel "lava"

|← —————————— 222 feet (69 m) —————————— →|

© Infobase Learning

The same RBMK reactor after it exploded and caught fire, sending the contents
of its core into the air over Europe

by simply having more neutrons in the reactor than the xenon-135 could absorb. Xenon-135 always reached an equilibrium concentration in the reactor, because it decayed quickly into a nonabsorbing isotope, cesium-135, due to a nine-hour half-life, but the extent of the concentration depended on the power level. At highest power, the xenon-135 level was highest, but it was compensated for by the control rods. At a lower level, there was always a lower level of xenon-135, and the control rod positions needed to compensate were about the same. However, if the

(continued on page 126)

THE GHOST VILLAGE OF PRYPIAT

More than 1,200 buses in a column several kilometers long arrived in Prypiat to evacuate the residents, who were advised to take three days' provisions with them. At 2:00 P.M. on April 27, 1986, about 44,600 people started climbing into the buses, and it took only three hours for everyone to be loaded and moved out. Evacuees were placed with families in adjoining districts beyond what would become the Chernobyl Exclusion Zone. By May 2, the radiation contamination situation was getting worse, and the evacuation zone was expanded to include everyone within a 19-mile (30-km) radius of the Chernobyl-4 reactor. It took four days to evacuate everyone. Although everyone was told that this was a temporary situation, the city of Prypiat remains contaminated with radioactivity, and the residents were never able to return. Just about everything they owned still sits where it was at 2:00 P.M. on April 27, 1986.

Prypiat was once considered the most beautiful city in the Ukraine. Everything in the city was new. It had been built specifically for the people working at the

Police must guard the entrance to Prypiat to prevent people from being exposed to radiation (AP Images)

large nuclear power complex in nearby Chernobyl, and its population was expected to grow to 80,000. Young people from 13 nationalities poured into the city, with an average age of 26 years. During construction, every native bush and tree was preserved and harmoniously incorporated into the new city's landscape. In the downtown area were built a cultural center, bookstore, hotel, four libraries, an art school with a concert hall, secondary and vocational schools, and a medical complex. Spacious apartments were built for everyone, and at the center of the town was an amusement park with a Ferris wheel. On the edge of town were acres of greenhouses.

Today, the city is empty of people. It is a forbidden zone. Plants are growing in an uninhibited manner, and grass is gradually taking over streets, walkways, and parking lots, as shrubs and trees take over buildings. The city of Prypiat is a study of what would become of all human-built civilization if everyone were to leave the planet. Windows break, and dirt blows into buildings. Plants start to grow on ledges and in corners, as classrooms and laboratories start to lose their freshly abandoned look. Looters have come, gone, and come again, but it is difficult to carry away every object in an entire city by hand, and most of everything remains in place. Hundreds of heavy-lift helicopters lie abandoned. In the Prypiat prison, all the cell doors are sprung open.

Radiation remaining in the city seems unevenly distributed. It is not as if radiation-bearing dust came straight down and settled on everything. The placement of building walls and even trees seemed to steer the radioactive dust in certain directions as it blew into the city, and subsequent rain and wind may have redistributed it. Some sections of the city are virtually radiation free, while some sections are as dangerous to walk in as they were 20 years ago.

Former residents were distributed into 7,700 apartments in nearby Kiev and Chernigov, and more than 21,000 farmhouses were built for evacuees later in 1986.

Although it was virtually unknown at the time, the city of Chernobyl-2 was located just northwest of Chernobyl, but it was never shown on a map. It was the site of a massive over-the-horizon radar installation, built in the 1970s to monitor incoming missiles from submarines in the Atlantic Ocean or from the continental United States. More than 1,000 people worked at the radar station, named Duga-1. The small town had one street, named Kurchatov. The radar antenna was enormous, several city blocks long and almost as high, and it transmitted a 10 megawatt signal at somewhere between seven and 19 megahertz. It contaminated shortwave radio listening all over the world for decades, and it was called the "Russian Woodpecker" for the sound it created. Today the radar facility is abandoned and is gradually being taken over by vegetation, just as is Prypiat.

(continued from page 123)

power were suddenly decreased, then the xenon-135 level would not be commensurate with the new power level, and the control rods would have to be pulled way out to compensate. The way to get around this was to reduce the power level very slowly, allowing the excess xenon-135 to naturally decay up to cesium-135 and establishing a new, lower equilibrium level. This phenomenon was well known to the experienced operators on the day shift.

The shift changed, and the night workers took over, with the reactor still running at high power. The night shift consisted of a crew with little or no experience in nuclear power, mostly drafted from nearby coal-fired plants without any special education. One of the operators had a little experience working on a nuclear submarine. They were handed the task of completing the tricky experiment with a simulated emergency shutdown. At 11:00 P.M., they got the call from the grid controller, giving them the go-ahead to reduce power to the test level, and they proceeded to drop the power to 1,000 thermal megawatts. Having no knowledge of xenon-135 neutron absorption, they reduced the power quickly, and they were surprised to find the fission going into an uncontrolled free fall, bottoming out at 30 megawatts, or about 5 percent of their target power.

The operators found this behavior strange, and they concluded that the automatic power regulator must have failed. They proceeded to pull out the controls, to try and compensate for the lack of power, pulling them out beyond the limits allowed under safety regulations. The level of xenon-135 produced by the high-power operation was difficult to overcome, and the crew was able to raise the power level to only 200 megawatts, or a third of what was necessary for the test. In spite of this failure to meet the test specifications, the operators decided to run the test anyway, and at 1:05 A.M. on April 26 they turned on the emergency water pumps that were supposed to be powered by the free-spinning turbogenerator. The emergency pump action added to the normal coolant pump action increased the coolant flow beyond safety regulations, and the resulting lack of any steam in the reactor core increased the amount of liquid water in the core. At 1:19 A.M., the unusual mass of water in the core degraded the fission activity to the point that the controls rods had to be completely removed from the core. The reactor was now in a very precarious situation, in which the only control was provided by xenon-135 and water.

At 1:23:04 A.M., the test was started by shutting off the steam line to the turbine. For the next 36 seconds, the test was affirmative, as the emergency pumps spun at full power from electricity made by the turbogenerator, still turning by momentum. However, the steam line to the turbine was shut for the safety test, and this caused steam voids to form quickly in the reactor. The moderation improved, and the reactor became supercritical. As the power level increased, the xenon-135 was burned off by capturing an enhanced concentration of neutrons, and with less xenon-135, the power increased ever faster.

By 1:23:40 A.M., the reactor was out of control and increasing in power. Someone in the control room hit the AZ-5 button, the manual scram. The control rods had been run up to the end-of-travel and even under emergency conditions the rods moved slowly. It took 18 to 20 seconds for the bank of rods to run completely into the core, and for the first few seconds the rod insertion was actually increasing the fission activity, instead of decreasing it. The power of the reactor spiked suddenly, and fuel rods began to fracture. The breaking fuel rods warped the control rod passages, and the controls jammed at about one-third in. The reaction continued, unabated. At 1:23:47 A.M., the power of the 3,200 megawatt reactor jumped to 30,000 megawatts. Fuel rods began to melt.

The coolant was flashed to superheated steam, and it had nowhere to go. The reactor exploded, sending the heavy top plate through the roof and carrying away the massive refueling crane. The graphite was white hot, and with the roof gone, fresh air rushed in. The graphite caught fire, sending the destroyed fuel and its one-year accumulation of fission products up into the atmosphere in a tall black cloud.

HEROISM, EVACUATION, AND A PROBLEM TOO BIG TO COVER UP

Radiation levels in the reactor building immediately climbed to 20,000 roentgens/hour (18,700 rem/hour, or 18.7 Sv/hour). A lethal radiation dose is 500 roentgens absorbed over a period of five hours. Unprotected workers received fatal doses within several minutes of exposure. All workers were equipped with dosimeters to monitor individual radiation doses and prevent overexposure, but the maximum reading was 1,000 roentgens, and the instruments quickly ran off the scale. Some radiation instruments failed immediately when turned on. The operations crew could determine only that the radiation level was somewhere above 3.6 roentgens per hour. The true reading was 5,600 times higher.

Seeing the relatively small radiation readout on the instruments, the reactor crew chief, Alexander Akimov, assumed that the reactor was intact and he was facing a minor problem. Down in the lower chambers of the reactor building, the missing top of the structure was not visible. The fact that graphite chunks and pieces of fuel were coming down in the yard was ignored, and Akimov and his crew stayed in the building all night, trying to pump cooling water into the reactor. Akimov and most of his crew died within three weeks of radiation sickness.

Seeing the telltale column of smoke, the Chernobyl Power Station firefighter brigade, under the command of Lieutenant Vladimir Pravik, arrived to extinguish the fire. Using high-pressure water, the men managed to extinguish the fire on the roof by 5:00 A.M., but they had not been warned that it was a reactor fire and the area was hot with extreme radiation. Pravik died on May 9, 1986. The graphite, far down in the building, burned for days, until it was finally put out by helicopters flying over and dropping material on the fire. They tried dropping sand, then lead, clay, and finally boron, to stop the fission process. The smoke contained massive quantities of uranium and all fission products, including the particularly dangerous isotopes cesium-137, iodine-131, and strontium-90.

A quickly formed government committee, led by Valeri Legasov, arrived at the scene of the accident that evening, April 26. By that time, two people were known dead and 52 were hospitalized with radiation poisoning symptoms. The workers reported a strange taste in the mouth, like metal, and a feeling of pins and needles on the skin. The destruction of the reactor was recognized and acknowledged. It was a Level 7 nuclear disaster.

On April 27, 1986, a worker showed up for the day shift at the Forsmark Nuclear Power Plant in Forsmark, Sweden, and walked through the radiation detection gate on his way into the reactor building. Everyone working at a nuclear plant was checked for radiation, going into or out of the building. Strangely, the radiation alarm went off as he entered. A quick check with a handheld instrument showed that his clothing was covered with radioactive dust. Where had he been?

He had been outside, walking into the plant. His clothing had picked up radioactive fallout, which was coming down in Sweden from the explosion and fire at Chernobyl, six hours before and 680 miles (1,094 km)

Damaged nuclear power reactor number 4 at Chernobyl in the Ukraine *(© Igor Kostin/Sygma/Corbis)*

away. Calls went out to other nuclear plants in northern Europe to check for atmospheric radiation. Those plants located in a straight line to the Ukraine found contamination. The reactor accident, which the authorities in the Soviet Union had hoped to keep secret, quickly became known internationally. There was no way to hide such a huge release of radiation.

Back at Chernobyl, the government committee ordered the temporary evacuation of the city of Prypiat, where plant workers and their families lived. Evacuation began promptly at 2:00 A.M., and people were told not to take baggage. They were assured that they would be returning as soon as the problem at the reactor was repaired. At Chernobyl-4, the fuel was still generating power, and it started to burn through the concrete floor. The high temperature in the building was made worse by the helicopters dropping material on top, and a mixture of melted concrete and uranium began to flow into the lower reaches of the massive building. Fearing another steam explosion, a team of two soldiers, with a worker to show them valve positions, donned wetsuits and ventured into the lower regions of the building to open the sluice gates. Their flashlight failed as they entered a room, and they had to find their way out by feeling along a pipe. All were exposed to fatal doses of radiation. In all, 55 workers died of direct exposure to radiation at Chernobyl.

Reactor 4 at Chernobyl, with a concrete cover temporarily controlling radiation release *(AP Images)*

The power plant was a total loss, and rather than try to tear it down and bury the remaining radioactive material, the entire building was encased in an enormous concrete sarcophagus. Work was completed by December 1986. The hundreds of vehicles used in the cleanup operation were too contaminated to be driven off-site, and they were parked in a field nearby. They remained there for many years. The sarcophagus is deteriorating, and an engineered barrier to cover it is being built.

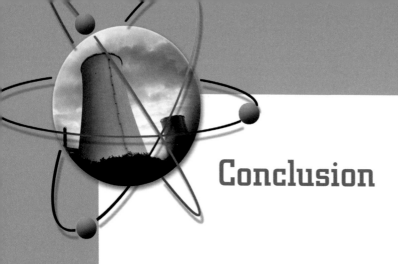

Conclusion

It is instructive to see nearly all the nuclear power disasters, accidents, incidents, and mistakes ever made compiled in one document. Patterns and trends become clearer, and the general maturing of nuclear science and engineering over the past 100 years can be traced, as it grew from an embryonic curiosity to a major source of fundamental power. There was an odd "jump" in the middle of the chronology section as nuclear research became a military and political point of force. Always danger-ous in new and strange ways, nuclear technology rushed forward at an unprecedented speed, but the concept of human safety was never far behind the steady push to expand knowledge and expertise in this spe-cialized field. It is only through detailed study of what has gone wrong in nuclear systems, nuclear experiments, and nuclear processes that safety can be improved. The research and development phase of nuclear tech-nology is not finished, but it may at least have reached a resting point, at which safety concerns can finally catch up and become the larger part of the nuclear art.

Nuclear fission as a discovered principle was a double-edged sword. It was a shortcut to releasing the enormous energy locked up in the atomic nucleus, and there were obvious applications for such a compact, intense source of power. With the power came a danger. In releasing such a large measure of energy in such a short period of time came radiation such as had never before been experienced, and as the energy extraction technology grew, safeguards for human life were an important side issue. Bad experiences with X-rays or radioisotopes had to be scaled up to a new level of health preservation and another couple of decades had been necessary for these lessons to be learned. In all, the following conclusions may be drawn from the chronology of nuclear power accidents:

❊ **Nuclear technology and economics do not mesh smoothly.**
   The purpose of nuclear technology is a controlled release of power from atomic nuclei, and the purpose of fiscal management is to accomplish this with as little expenditure of money as possible. So far, nuclear technology has been extremely expensive to implement, and this is at cross-purposes with economics. Nuclear power technology is exotic and costly to make safe.

   A premier example of the effects of fiduciary responsibility was the fire in the Windscale Unit 1 reactor in the United Kingdom in 1957. This was a reactor built with a budget made as lean as possible. Great Britain saw a very urgent need for a plutonium production reactor to come up to speed with the United States, who had just spent billions of war dollars constructing a plutonium production facility the size of Rhode Island in a desert. The British government lacked the funds, the facilities, and the natural resources for such an ambitious development, but a looming cold war seemed to make it necessary. A minimal facility was thrown together with available manpower and materials, and it met the minimum requirement, to produce enough plutonium-239 under maximum operation to build a single atomic bomb.

   The specifications were met within the necessary budget, but at an ultimate cost. The Windscale-1 production reactor caught fire. A nuclear fire is unlike any other industrial fire. It does not simply destroy the production facility; it also has the potential of destroying everything surrounding the production facility with randomly scattered radioisotopes. Great Britain exists on a fairly small island, and everything surrounding a production facility, depending on the affected radius, could have been a large percentage of the entire kingdom. The British nuclear industry was fortunate that the accident was confinable, but the implications of this cheaply developed facility and its inevitable breakdown are profound. Five decades later, this accident is still being cleaned up. Inadequate engineering, experimentation, and data gathering had been applied to the reactor construction. In the middle of the building project, when it was too late to reverse course for time and money considerations, a modification, the "Cockcroft follies," was added, and this jury-rigged modification was all that saved the island.

The unusual level of expense for safety-related nuclear technology continued to restrict the industry throughout the last 50 years, and in the 1970s economic realities at least stalled the push for nuclear power, though not disbanding it entirely. The great Chernobyl disaster, which also curtailed the advancement of nuclear power, was primarily due to economic considerations. Money was saved by not training reactor operators. A badly designed reactor was built anyway, because of the time and money required to design another one. Military plutonium production and civilian power production were combined into one enterprise, for cost savings. In the broadest sense, economic measures have historically resulted in badly planned, badly developed reactor projects, and a badly designed reactor is a dangerous reactor.

❋ **On the frontier, there will be problems.**
Early nuclear technology suffered from being constantly on the frontiers of science and engineering, and there are always unknown dangers and pitfalls on the leading edge. For this reason, and for security measures, all early nuclear reactor experiments in the United States were carried out in remote, uninhabited locations. It was a habit to literally locate a new, untried reactor in a place where, if it exploded, it would not affect a stationary population. The first breeder reactor was therefore built in the desert in the middle of Idaho, nuclear rocket engines were tested at Jackass Flats, Nevada, and nuclear aircraft components were tested in a remote, rural setting in North Georgia.

The fact that these early efforts were prone to meltdowns, explosions, and radiation releases was, in fact, predictable, and it was appropriate to locate them away from civilization. There were, and still are, many unknowns in nuclear engineering in the middle of the 20th century, and to expect such a new undertaking to be pushed forward without incident would not be clear thinking. Nuclear fission was dangerous in ways that had never been considered, and some accident modes were not entirely predicted. In the sodium reactor experiment at Santa Susana, California, in 1959, a problem with an exotic bearing coolant was not predictable, and the meltdown it caused was not even detected for 13 days after it happened because no one was looking for it. The frontier is, by its very nature, dangerous. A heightened vigilance is necessary, and each incident must be entered into the common store of knowledge.

The abandoned city of Prypiat, with reactor 4 at Chernobyl in the background *(Yuri Kozyrev/ Newsmakers/Getty Images)*

✤ **A worker cannot be overeducated.**

Among the most astonishing modes of nuclear accident is the criticality accident in fuel processing. By some stuck valve or misplaced container, a mass of fissile material, usually dissolved in water, is brought together accidentally, and it becomes a supercritical nuclear reactor. In this configuration, the solution of nuclear fuel produces the extreme radiation flux that is usually confined to heavily shielded reactor vessels, and anyone near will probably die. The solution to this potential problem seems utterly simple. In the design and construction of a nuclear fuel–processing facility, regardless of how complex or elementary the process, there should be no ability for liquid to collect in a size or shape capable of fission. The perfect shape for a fission reaction is a sphere, followed by a cylinder, the shape of a common soup can. The maximum allowable size for a given process can be calculated. Depending on the isotopes being processed, a fissioning reactor can be as small as a grapefruit. Container shapes that discourage fission are typically long, thin pipes or shallow trays.

Some of these incidents, as detailed in this volume, were admittedly too extraordinary to have been prevented, but the most tragic accidents occurred because the workers involved had no grasp of the significance of a critical mass when working with fissile materials, and there is no excuse for this. In a typical industrial plant, worker training may be minimal. There is no real need for a drill-press operator to know the theory of polyphase electric motors, even though one is atop his drill. In a nuclear fuel plant, on the other hand, every worker who may touch a valve must be schooled in the cause and effects of nuclear criticality. A worker who fills bottles with fluid or transfers metal billets from box to box has no intuitive concern for the consequences of his changing the procedure slightly. The most unskilled worker in such a plant must have the training and the comprehension of an undergraduate physicist in this narrow field. A perfect example of this problem is the Tokaimura criticality accident in Japan in 1999, and the Chernobyl disaster would not have happened if the reactor operations crew had possessed an elementary understanding of fission dynamics.

※ **It is inappropriate to make safety findings secret.**

If an unexpected nuclear accident has occurred, then the only way to ensure that it may never happen again, anywhere, is to at least publish an account of what happened. In the past, such an open exchange of accident data was rare. In early manufacturing situations, such as the radium dial plants, the sickness and death of workers due to radium ingestion was covered up as a top industrial secret. The reasons for the secrecy were to prevent lawsuits and to keep worker salaries low, but as a policy this secrecy led to dozens if not hundreds of needless deaths in the radium industry. So bad was this covering up of radiation sickness, it eventually destroyed the industry and resulted in federal laws against the use of industrial radium. The radium industry ruined itself with its well-enforced secrecy policies.

The atomic bomb projects of World War II were rigidly enforced closed-shop operations, often with a minimum of safety-related information interchange. The deaths of two scientists at the Los Alamos atomic bomb development laboratory in 1945, for example, were kept secret and were known only as rumors to workers in the same workplace. Elaborate cover stories were concocted to explain the deaths. While the stories did not deceive the scientists, they still had no clear, official word of what exactly had happened. Eight months later, the same apparatus killed yet another scientist at the same laboratory for the same reasons. If at least the bomb research community had known the correct details, it is likely that the deadly experiment mode would have been heavily modified.

For decades, nuclear fuel–processing plants did not share information concerning criticality accidents involving water solutions of fuel compounds. One reason for the continuing occurrences of this type of accident was that no one heard of what was happening at other facilities. A fine counterexample of this phenomenon was the way the Atomic Energy Commission treated the SL-1 explosion in Idaho. This incident was considered to be a fundamental breakdown of the engineering and implementation of nuclear power. From a purely technical standpoint, it was the worst nuclear reactor accident in U.S. history, and its analysis was treated with a thorough investigation. News of the event and results of the investigation were made public, at least to the nuclear engineering community. Posters were made, and a motion picture of the accident and the cleanup operations was produced. For decades, this accident was studied in engineering classrooms and cited as an example of how not to design a nuclear plant. Treatment of the SL-1 accident was an example of how all nuclear incidents should have been handled from the beginning. There was never another exploding boiling water reactor, partly because of engineering changes and an enhanced awareness of dangerous procedures.

✳ **There is always the human factor.**
No matter how well educated are the nuclear operators, technicians, managers, engineers, and scientists, there remains the possibility for mistakes to be made. Personnel can be properly and extensively trained, well paid, and working in a well-kept, well-designed environment, but still costly or deadly errors can be made. A lack of sleep the previous evening, a touch of the flu, a domestic problem, or simply a slip off into daydreams can cause a wrong move and bring an emergency crashing down. A severe marital problem may have caused one of the worst nuclear power accidents, the SL-1 explosion. It would seem that nuclear power plants and fuel-processing facilities should be designed so that no single error can cause a major breakdown. There have been major efforts in this direction, but sometimes several small errors add up to a catastrophe. In case after case, particularly in Soviet weapons research laboratories and fuel-processing plants, otherwise knowledgeable workers and scientists wanting to shortcut a procedure for the sake of saving time caused fatal accidents. A human tendency to find shortcuts and a natural impatience are not well placed in nuclear work.

There is also a human tendency to want to find the cause of a problem quickly and resolve it before it grows into a bigger problem. This was the main human problem at Three Mile Island in 1979. Operators were too analytical for their own good, trying to figure out what was going wrong in the reactor system and jumping to correct what they perceived as causes of the shutdown. If only they had kept their hands off the switch panels for a while, the automatic logic circuits built into the machinery would have detected a stuck-open valve and would have correctly directed cooling to the reactor core. It required a lot of additional training and procedure modifications to dissuade operating staffs in power plants from this behavior. This is a difficult factor to eliminate, and it probably still exists, buried in the subconscious of all who operate nuclear equipment.

These are only some of the most obvious conclusions to be drawn from a continuing study of the complexities of nuclear power accidents and disasters. Nuclear power incidents, such as those detailed in this volume, will be analyzed and contemplated in detail by engineers and scientists as the world moves forward in the quest for a clean, reliable power source.

1879   In November, in a deep mountain canyon in Barry County, Missouri, Bill Henry, John Dempsey, and Bill Boyceyer accidentally discover a vein of what appears to be rich silver ore in a cave. The three men experience acute radiation poisoning, and it would turn out that their find was an unusual concentration of natural radiation.

1896   On December 1, Nikola Tesla, an inventor headquartered in New York, writes warnings of the biological hazards of working with radiation in *Electrical Review,* after his own experiments with X-rays. In exposing his hand to X-rays, he experiences a painful irritation of the skin, inflammation, the appearance of blisters, and open sores.

1904   In October, Clarence Madison Dally, a glassblower at the Edison Laboratory in West Orange, New Jersey, dies of mediastinal cancer, the cause of which is obviously his work with X-rays. His face and hands had been exposed to X-rays while holding his hand in a beam and gazing at it with a fluoroscope.

1929   On August 14, Peg Looney, a worker at the Radium Dial factory in Orange, New Jersey, dies of radiation poisoning.

1932   On March 31, Eben McBurney Byers, an American socialite and industrialist, dies a gruesome death caused by ingesting a popular medicine containing radium.

1934   On July 4, Marie Curie, codiscoverer of radium, dies of aplastic pernicious anemia, caused by her long exposure to the radiation from radium and X-rays.

Late 1940s The ZEEP assembly at Chalk River, Ontario, goes unexpectedly critical because a technician inserts a chip of wood into a pump-control button to save time. The resulting burst of radiation scrams the NRX reactor in an adjacent building, and three people are seriously exposed to radiation. This incident was unreported.

1952   On June 2, at Argonne National Laboratories in Illinois, operating procedures are violated and an experimental reactor goes above prompt critical, ruining the core and giving four people significant radiation exposures.

On December 12, the heavy water–moderated NRX reactor at Chalk River, Ontario, Canada, undergoes a partial core meltdown due to operator error. This incident has the distinction of being the world's first recorded reactor core meltdown.

1953   On March 15, at the Mayak Production Association in Russia, too much plutonium nitrate solution in one vessel goes critical. One worker is seriously exposed to radiation, and one is significantly exposed.

On April 9, at the Arzamas-16 secret nuclear weapons laboratory in the Soviet Union, an error in the setup wrecks a "vertical split table," which is a remote-control device for experimenting with bomb core assembly. It eliminates the dangers of manual assembly that were so deadly at Los Alamos, but there is still room for improvement of the procedure.

1955   On November 29, the EBR-1 experimental breeder reactor at the National Reactor Testing Station near Arco, Idaho, has a partial fuel meltdown due to operator error.

1956   On March 17, Irène Joliot-Curie, daughter of Marie Curie, dies of leukemia. The leukemia was certainly caused by long-term exposure to radiation in the laboratory, but the fatal dose probably occurred in 1946, when a vial of highly radioactive polonium exploded in her face.

On July 3, an unexpected prompt criticality occurs in a remotely controlled, experimental honeycomb assembly of uranium and graphite at Los Alamos. This is considered a "dragon-type" excursion, and nobody is harmed, due to the fact that everyone was one-quarter mile away. This was the first incident of lives being saved because of lessons learned from previous deaths at Los Alamos.

1957   On April 27, at the Mayak Production Association in Russia, uranium precipitate builds up in a filtrate-receiving vessel, and it goes critical. One person dies, and five are injured.

On September 29, a large tank of radioactive waste explodes at the Mayak plant when the cooling system fails. At least 200 people die of radiation poisoning, 470,000 are exposed to injurious levels of radiation, and 10,000 people are evacuated.

On October 10, the graphite core of Windscale pile number 1 catches fire at Sellafield, in Great Britain, sending a great deal of radioactive smoke up the stack and contaminating the surrounding area.

1958   On January 2, at the Mayak Production Association in Russia, too much uranyl nitrate solution winds up in one vessel, and it goes critical. The prompt burst kills three people, and one is injured.

On May 24, a metallic uranium fuel rod catches fire and is torn in two as it is being removed from the core at the NRU reactor at Chalk River, Ontario.

On June 16, at the Oak Ridge, Tennessee, Y-12 plant, uranyl nitrate winds up in a water collection drum. It goes critical several times, significantly exposing seven people to radiation.

On October 15, an accidental high-power excursion takes place at the heavy water–moderated research reactor in the Boris Kidrich Institute in Vinča, Yugoslavia, due to saturated radiation detectors giving false readings. Six scientists receive high radiation doses.

On November 18, the High Temperature Reactor Experiment goes wild for about 20 seconds at the National Reactor Testing Station, due to an instrument failure. This accident is unique, as it is caused solely by instrumentation. The core is melted.

On December 30, at the Los Alamos Scientific Laboratory in New Mexico, too much plutonium organic solution winds up in a tank. It goes critical, and one person is killed. Two more are significantly exposed to radiation.

1959   On July 13, the Sodium Reactor Experiment experiences an inexplicable power excursion that causes a partial core meltdown. The melt was not discovered until 13 days later. Radioactive gases are released.

On October 16, at the Idaho Chemical Processing Plant, uranyl nitrate solution winds up in a waste-receiving tank. It goes critical multiple times, and two people receive significant radiation exposure.

1961   On January 3, the SL-1 stationary low-power reactor near Idaho Falls, Idaho, goes prompt critical during maintenance activities and explodes, killing the three operators.

On July 14, at the Siberian Chemical Combine in Russia, uranium hexafluoride gas accumulates in a vacuum pump oil reservoir. It goes critical twice, and one person receives a significant radiation dose.

1962   On September 7, at the accident-prone Mayak Production Association in Russia, plutonium nitrate solution builds up in a dissociation vessel. It goes critical three times, but there are no radiation exposures.

On December 11, the crew at Los Alamos assumes that a zero-power critical assembly had been run successfully the previous day and proceeds to run an experiment. The assembly had not been successfully run the previous day, and the zero-power core starts generating 200 watts. They try to shut it down, but it goes prompt critical, producing a megawatt. Crew is embarrassed.

1963   On January 30, at the Siberian Chemical Combine in Russia, too much uranyl nitrate solution in a collection vessel goes critical several times, but nobody is exposed to radiation.

On March 26, at Lawrence Livermore Laboratory in California, concentric cylinders of uranium suddenly become highly supercritical, catch fire, melt, and spread molten uranium on the floor. The uranium cylinder at the center of the assembly was slightly off center. It is remotely controlled, so nobody is injured.

On December 2, at the Siberian Chemical Combine in Russia, too much uranium organic solution in a vacuum system holding tank goes critical several times, but there are no reported personnel exposures to radiation.

1964   On July 24, at the United Nuclear Fuels Recovery Plant at Wood River Junction, Rhode Island, uranyl nitrate solution in a carbonate reagent makeup vessel goes critical twice. One person is killed, and two receive significant radiation exposures.

1965   On May 16, the SNAP-10A 500 watt nuclear reactor is irretrievably scrammed in outer space due to a stuck relay.

On May 28, at White Sands Missile Range in New Mexico another Godiva reactor goes unexpectedly supercritical, snapping off the heads of the assembly bolts and tossing parts. The original assembly bolts had been stainless steel, but they had been replaced with bolts made of uranium. A safety interlock is bypassed, and trouble results.

On November 3, at the Electrostal Machine Building Plant in Russia, uranium oxide slurry collects in a vacuum vessel and goes critical. There are no radiation exposures.

On December 16, at the Mayak Production Association in Russia, uranyl nitrate solution collects in a dissolution vessel to critical size. There are multiple power excursions, but no reportable radiation exposures.

On December 30, at the VENUS assembly, which is an experimental heavy water reactor in Mol, Belgium, an operator standing over it manipulates the manual controls in the wrong sequence, and he gets a heavy dose of radiation as the reactor goes critical. His left foot receives a massive 4,000-rem dose and must be amputated.

1966 On October 5, a blockage in the cooling channels causes a partial core meltdown in the Fermi 1 fast breeder reactor in Monroe County, Michigan. Analysis and reduction of the damage is difficult, due to a lack of preparation and anticipation of such incidents.

1967 In May, the Chapelcross Nuclear Power Station at Dumfries and Galloway, Scotland, has a partial meltdown and fuel fire. Contamination is limited to the reactor core.

1968 On December 10, at the Mayak Production Association in Russia, aqueous and organic plutonium solutions are allowed to collect in a 60-liter vessel. Three power excursions result, killing one person and injuring another.

1969 On January 21, an experimental reactor at Luccens, Canton of Vaud, Switzerland, explodes. The reactor is in an underground cavern, which contains the accident and is sealed off.

1971 On February 15, experiments to evaluate the effectiveness of iron and beryllium in experimental reactor SF-7 at the Kurchatov Institute in Russia go awry due to poor estimates of criticality and a lack of necessary calculations. The reactor goes unexpectedly prompt critical and pulses 50 times, seriously injuring two people, with both receiving a dose of 1,500 rem to their feet.

On May 26, again at the Kurchatov Institute in Russia, as experimental reactor SF-3 is being used to measure the critical mass of some highly enriched uranium fuel rods, the water is drained from the reactor tank and the fuel matrix collapses from its own weight and it goes heavily supercritical. Nobody is harmed, but the engineering blunder is large.

1975 On March 22, in unit 1 at the Brown's Ferry Nuclear Power Station near Decatur, Alabama, a worker using a candle for an air leak at the cable penetration seal between two rooms manages to set the plant on fire. This accident led to an addition to the Code of Federal Regulations, 10 CFR 50.48 concerning fire protection in nuclear plants.

1977   On February 22, a fuel assembly is damaged by extensive corrosion in the KS 150 reactor at power plant A-1 in Jaslovské Bohunice, Czechoslovakia. Workers had failed to remove moisture-absorbing material from an assembly before loading it into the reactor. The reactor was a total loss.

1978   On October 17, at the Idaho Chemical Processing Plant, a uranyl nitrate solution in a lower disengagement section of a scrubbing column goes critical. There are no injuries.

On December 13, at the Siberian Chemical Combine, plutonium metal ingots are placed in a storage container. It goes critical, and eight people are exposed to radiation.

1979   On March 28, unit 2 at the Three Mile Island Nuclear Generating Station near Harrisburg, Pennsylvania, experiences a partial core meltdown due to multiple system failures. It is the worst commercial power plant incident in the United States. The reactor is a total loss.

1980   On March 13, reactor A2 at the Saint-Laurent Nuclear Power Plant in Orléans, France, has a brief power excursion, wrecking a fuel bundle and sending some minor radioactivity into the surroundings. The reactor was repaired and continued operation.

1981   In March, more than 100 workers at a nuclear power plant in Tsuruga, Japan, are exposed to more radiation than they are allowed in the company's procedures manual.

1983   On September 23, at the RA-2 facility experimental test reactor in Buenos Aires, Argentina, the operator is killed by radiation exposure due to an error while reconfiguring a fuel plate. The reactor is supposed to be shut down, but he manages accidentally to make it supercritical.

1986   On May 12, a cooling tower at the Brown's Ferry Nuclear Power Station near Decatur, Alabama, having dried out from being idle ever since the cable fire back in 1975, catches fire and burns to the ground. Under normal operation, the cooling tower, made of redwood, fiberglass, and aluminum, is kept constantly wet with a water cascade.

On April 26, reactor number 4 at the Chernobyl Plant, near Prypiat in the Ukraine, explodes and burns due to a badly conducted safety system test. It is the worst reactor accident in world history. Prypiat was permanently evacuated.

On May 2, in an experimental THTR-300 high-temperature gas reactor in Hamm-Uentrop, Germany, a fuel pebble lodges in a fuel-delivery pipe. Attempts to break it loose damage the pebble, releasing fission products detectable two kilometers from the reactor.

1989   On November 24, a poorly built Soviet reactor damages 10 fuel rods in Greifswald, East Germany, while operators are testing an emergency safety system.

1993   On April 6, in a concrete bunker under Building 201 at the Tomsk-7 Siberian Chemical Enterprise plutonium reprocessing facility in Tomsk, Russia, an accidental pressure buildup leads to an explosive failure in a 34-cubic-meter reaction vessel. Two thousand cleanup workers and 160 on-site personnel are subjected to unusually high levels of radiation exposure.

1997   On May 15, at the Novosibirsk Chemical Concentration Plant in Russia, uranium oxide slurry and crust collect in the lower regions of two parallel vessels, and the combination of the two goes critical multiple times. Nobody is injured.

1999   In June, at the number 1 reactor of the Shika Nuclear Power Plant in Ishikawa Prefecture, Japan, an operator error leads to a 15-minute uncontrolled supercritical reaction. Attempting to insert one control rod, three are withdrawn instead. This incident is covered up with falsified records until March 2007.

On September 30, in the Tokaimura uranium reprocessing facility at Ibaraki Prefecture, Japan, workers mistakenly pour about 585 ounces (ca. 16.6 kg) of uranyl nitrate into a precipitation tank. The amount of uranium in the tank exceeds a critical mass, and the container goes supercritical. Two workers die of radiation exposure.

2003   On April 10, a spent fuel rod being cleaned in a tank of cold heavy water breaks open and spills fuel pellets at the Paks Nuclear Plant in Paks, Hungary.

2005   On April 19, a great deal of uranium and plutonium dissolve in around 21,926 gallons (83,000 l) of nitric acid leaked from a cracked pipe and into the sump at the Thorp Nuclear Fuel Reprocessing Plant at Sellafield in Great Britain.

2006   On March 6, highly enriched uranium solution leaks during transfer into a lab at the Nuclear Fuel Services Erwin Plant at Erwin, Tennessee. The incident causes a seven-month shutdown of the plant.

2008   On December 22, a massive uncontrolled release of radioactive material has nothing to do with nuclear power. The ash pond at the Tennessee Valley Authority's Kingston coal-fired power plant breaks during a rainstorm, sending 20 curies of naturally occurring uranium in the ash covering the surrounding territory.

2011   March 11, Japan is hit with its biggest earthquake in history, followed quickly by a powerful tsunami. The Fukushima I nuclear power plant, with six boiling water reactors, experiences core meltdowns, hydrogen explosions, and is damaged beyond repair when its diesel backup generators are flooded with seawater.

activation   making a substance artificially radioactive by bombarding it with neutrons

alpha particle   also alpha radiation, a class of ionizing radiation composed of a helium nucleus traveling at high speed

Army Nuclear Power Program   a program by the U.S. Army to develop small nuclear reactors for use in remote, inaccessible locations, headquartered in Fort Belvoir, Virginia

atomic energy   an antiquated term meaning energy that is released by the fission of heavy nuclei or the fusion of light nuclei

becquerel   or Bq, the SI unit of radiation measurement, equaling 2.7×10^{-11} curies

beta radiation   or beta particle, either an electron or a positron ejected from a decaying nucleus

breeder reactor   a nuclear reactor that makes more fuel, through neutron capture in non-fissile nuclei, than it uses to produce power

Browns Ferry Nuclear Power Station   a nuclear power plant having three General Electric boiling water reactors, owned by the Tennessee Valley Authority and located on the Tennessee River, near Decatur, Alabama

BWR   boiling water reactor, a single-loop reactor in which light water is used as the moderator and the coolant

chain reaction   a series of chemical or nuclear reactions in which each reaction causes another reaction

Chapelcross Nuclear Station   a nuclear power station in the town of Annan in Dumfries, Scotland, consisting of four MAGNOX gas-cooled graphite reactors

Chernobyl   a town in the Ukraine and site of the worst nuclear disaster in history in 1986

cobalt-60   a radioactive isotope of cobalt, emitting two powerful gamma rays at 1.2 and 1.4 MeV with a half-life of 5.27 years

control rod   a metal rod made of a neutron-absorbing metal, such as cadmium, used to soak up excess neutrons in a nuclear reactor and bring it to perfect criticality by adjustment

criticality   the balance state of a nuclear chain reaction, in which the number of neutrons being lost through leakage, unproductive capture, or fissioning capture exactly equals the number of neutrons being produced by fission

critical mass   the effective mass of uranium or plutonium fuel at which a nuclear reactor is critical

curie   a unit of measure of radioactivity, being 3.7×10^{10} nuclear disintegrations per second, or the amount of radiation coming from one gram of radium

decay chain   the series of steps through which a radioactive isotope progresses, becoming different and successively lighter isotopes, as it decays toward a final, stable, nonradioactive isotope

deuterium   heavy hydrogen, with the nucleus containing both a neutron and a proton

DOE   the U.S. Department of Energy

EBR   experimental breeder reactor, built at Arco, Idaho, in 1950

Elektrostal   Russia's largest center of fuel rod production for nuclear power plants, located 58 kilometers east of Moscow

enriched uranium   uranium reactor fuel that has had the U-235 content improved

Fermi 1   the world's first fast breeder commercial power reactor, in Monroe County, Michigan

fissile   a descriptor for an element that will release energy and excess neutrons when fissioned

fission   the splitting of a heavy nucleus into two lighter nuclei

fissionable   a descriptor for an element that can be fissioned by neutron capture

fission products   the lighter always radioactive isotopes into which a fissile fuel breaks upon fission

fluorescence   an energy conversion event at the atomic level, causing a visible glow

fuel reprocessing   the chemical separation of various components of spent nuclear fuel

gamma radiation   a high-energy electromagnetic wave, above X-rays on the electromagnetic energy spectrum, originating in the nucleus

Geiger counter   or Geiger-Mueller counter, an electronic radiation detector used to measure the presence of gamma or beta rays

GLEEP   the graphite low energy experimental pile, built at Harwell, England, in 1947

gray   or Gy, the SI unit of energy absorbed due to radiation exposure, equaling 100 rads

half-life   the time required for a radioactive sample to decrease its level of radioactivity by one half

heavy water   deuterium oxide, or water made with two deuterium atoms and one oxygen in each molecule

ionizing radiation   radiation of sufficient power to knock the top electron out of an atom upon collision, such as gamma rays, beta rays, and alpha rays

isotope   a subspecies of an element, distinguished by the number of neutrons in the nucleus

light water   ordinary water, as is available from a municipal tap

MAGNOX   a now obsolete British reactor fuel formula having magnesium cladding over uranium

Mayak   a nuclear fuel reprocessing plant in Russia, built in 1948, northwest of Chelyabinsk

Mev   a million electron volts, as a unit of energy applied to subatomic or subnuclear particles in motion

moderator   any substance used in a nuclear reactor to slow high-speed neutrons from fission down to thermal speed

neutron   a fundamental particle of matter having no electrical charge

Novosibirsk   Russia's third largest city, located in the Siberian Federal District

NRC   the U.S. Nuclear Regulatory Commission, an independent fee-for-services federal agency charged with the regulation and oversight of civilian nuclear activities involving source, by-product, and special nuclear material in the United States

NRTS   the National Reactor Testing Station, set up in the desert of Idaho, near Arco, after World War II

NRU   National Research Universal reactor, located in Chalk River, Canada

NRX   an experimental heavy water–moderated reactor built in Canada at Chalk River in 1947

nuclear physics   the study of forces, objects, and fields involved in the atomic nucleus and its interactions with matter

nucleus   the massive center of an atom, built of protons and, in all but one case, neutrons

Oak Ridge National Laboratories   a research, development, and production facility set up in a valley near Oak Ridge, Tennessee, during World War II for the atomic bomb project

Obninsk   a city in Kaluga Oblast, Russia, and site of the world's first nuclear power–generating plant

pile   an obsolete term, meaning nuclear reactor

plutonium   element number 94 in the table of the elements, artificially made by activating U-238

Prypiat   a modern city built in 1972 in the northern Ukraine, Kiev Oblast, and once the home of workers at the Chernobyl Nuclear Power Station

Pu-239   a fissile isotope of plutonium

PWR   pressurized water reactor

radioactive   capable of emitting radiation at a predictable rate by nuclear decay

rad   a unit of absorbed radiation dose, or enough radiation to deposit 100 ergs of energy in one cubic centimeter of water

radiation   a class of energy transmission by electromagnetic waves or by direct particle transfer

radioactive decay   the tendency of certain isotopes to undergo change in the nucleus

radioactivity   the emission of radiation, either by the willful manipulation of a nucleus or by spontaneous nuclear decay

RBMK   a now obsolete power reactor design using graphite moderator and boiling water coolant in the former Soviet Union

reactor   a machine or system built to sustain a neutron chain reaction in a fissile material for the purpose of power production

rem   roentgen equivalent man, a unit of radiation dose adjusted for its effectiveness against human flesh

roentgen   a unit of radiation measurement, or ionizing radiation enough to completely ionize one cubic centimeter of dry air at standard temperature and pressure

Santa Susana Field Laboratory   a rocket and nuclear reactor test facility in Moorpark, California, the site of the first commercial nuclear power reactor in the United States

Sellafield   the site of British nuclear power and weapons development, near the village of Seascale on the coast of the Irish Sea in Cumbria

SI   Système International d'Unités, the international metric system of measurements, headquartered in Paris, France

sievert   or Sv, the SI measurement of radiation dose, equaling 100 rem

thermal speed   the speed at which air molecules move at room temperature as they bounce around and hit each other

thermoelectric generator   an electrical generator that has no moving parts, fuel, or exhaust, using a self-heating mass of a radioactive isotope or a nuclear reactor

Three Mile Island   an island in the Susquehanna River in Pennsylvania, near Harrisburg, and location of the Three Mile Island Nuclear Generating Station

TMI   Three Mile Island

Tokai   a village in the Naka District, Ibaraki, Japan, home to the JCO fuel reprocessing plant

Tomsk   a city on the Tom River in the southwest of the Siberian Federal District of Russia, near the town of Severk

tritium   the heaviest isotope of hydrogen, having two neutrons in the nucleus

U-235   a fissile isotope of uranium, having 92 protons and 143 neutrons in its nucleus

U-238   a non-fissile isotope of uranium, having 92 protons and 146 neutrons in its nucleus

uranium   element number 92 in the table of the elements

Windscale   a pair of air-cooled graphite reactors built by the British government in 1946 to convert U-238 to Pu-239

X-ray   an electromagnetic wave above light in the electromagnetic energy spectrum originating in the electron cloud surrounding a nucleus in an atom

Further Resources

Arnold, Lorna. *Windscale 1957*: *Anatomy of a Nuclear Accident.* New York: Palgrave McMillan, 2007. A rare look inside the disastrous Windscale unit 2 fire in Great Britain. This important incident was not fully disclosed to the public until the publication of this book.

Fuller, John G. *We Almost Lost Detroit.* New York: Ballantine Books, 1975. Somewhat sensational, but still a good introduction to nuclear accidents in Canada, Britain, and the United States before the Three Mile Island incident.

McKeown, William. *Idaho Falls: The Untold Story of America's First Nuclear Accident.* Toronto, Canada: ECW Press, 2003. A wealth of detailed information concerning the SL-1 explosion, the only fatal nuclear reactor accident in the United States.

McLaughlin, Thomas P., Shean P. Monahan, and Norman L. Pruvost. *A Review of Criticality Accidents: 2000 Revision.* Los Alamos, New Mexico: USDOE, 2000. This is the premier resource on all criticality accidents, using unprecedented access to Russian records from their nuclear weapons program. Before the publication of this document, most of this information was classified SECRET.

Stacy, Susan. *Proving the Principle.* Idaho Falls, Idaho: USDOE, 2000. This is the official, DOE-sanctioned history of the Idaho National Energy Laboratory and the National Reactor Testing Station. It covers all of the nonsecret activities, including the SL-1 accident and the BORAX-I explosion.

Tucker, Todd. *Atomic America: How a Deadly Explosion and a Feared Admiral Changed the Course of Nuclear History.* New York: Free Press, 2009. An alternate view of the SL-1 explosion, with some further details. Admiral Rickover's submarine reactor was being tested a few miles away from the accident, and the interesting intersection of Rickover, SL-1, and the nuclear power industry is detailed in this book.

RECOMMENDED WEB SITES

Further depth on many topics covered in this book has become available on the World Wide Web. Available are detailed Web sites for any government facility or agency, some giving access to archives and histories. Any book listed above can be bought at www.amazon.com, and there are even entire books concerning nuclear power issues, long out of print or published by the Government Printing Office, that are available online.

Department of Energy Hanford Site is the official Web site of the Hanford facility, which was the production point for plutonium during World War II. The Hanford Site is still in operation, and this complete Web site contains everything from online videos to an abbreviations and acronyms directory. Although no longer producing plutonium, the Hanford Site is undergoing a massive environmental cleanup. Hanford's history also includes the Fast Flux Test Facility, an account of which is covered in some detail. Available online. URL: http://www.hanford.gov. Accessed January 18, 2011.

Google Earth, a utility program available free from Google.com, is an excellent way to view the sites of old reactor accidents. You do not need latitude/longitude coordinates to find many interesting sites. The site of the SL-1 explosion in the desert in Idaho, for example, may be accessed by simply typing "SL-1" into the "Fly to" box. There is no trace of the reactor left, but it is easy to see where the desert floor has been disturbed, and a short distance up the road from the clearing where the reactor stood, find where it is buried. "Sellafield" will take you to the site of the Windscale fire, but unfortunately the smokestacks have been torn down, so it is hard to pick out the existing Windscale reactors. Type "Chernobyl-4," and Google Earth will put you on top of the concrete sarcophagus that now covers the Chernobyl reactor. All the other units have been torn down. Find the main road, going northwest, and follow it a short distance to Prypiat, where you can see the entire ghost town, down to the standing Ferris wheel in the park. To use Google Earth, you must download the program into your computer system. It is free, and versions are available for the Mac, the PC, and for UNIX. Available online. URL: http://www.google.com/earth/index.html. Accessed January 18, 2011.

LANL Research Library is a vast collection of ebooks, databases, and news concerning all issues of nuclear science, nuclear technology, the history of nuclear topics, and current research and development. Students of nuclear science will find it an invaluable resource of information. The library search can find books, journals, patents, reports, videos, audiotapes, and recommended Web sites. Available online. URL: http://library.lanl.gov/. Accessed January 18, 2011.

Oak Ridge Associated Universities has a full copy of *A Review of Criticality Accidents: 2000 Revision* that can be read online, downloaded, or printed. This valuable reference is now in the public domain. Available online.

URL: http://www.orau.org/ptp/Library/accidents/la-13638.pdf. Accessed January 18, 2011.

Proving the Principle, an excellent history of the Idaho National Energy Laboratory by Susan M. Stacy, may be downloaded or read online. Available online. URL: http://www.inl.gov/proving-the-principle/. Accessed January 18, 2011.

Prypiat, the nuclear ghost town in the Ukraine, can be explored at many locations on the Internet. There are hundreds of ground-level photos of the decaying city at this official site. Available online. URL: http://pripyat. com/en/photo_gallery/pripyat/. Accessed January 18, 2011.

Sellafield Ltd gives a good insight into the involvement and attitudes of Europe in the worldwide nuclear industry. Although it covers only one nuclear facility in England, it is a good start in understanding the participation of Europe in the long-term energy solution of nuclear power. It includes the history of Sellafield, with a video of its construction, to the current uses of the facilities at Sellafield and the decommissioning of the Calder Hall reactors. Available online. URL: http://www.sellafieldsites. com/. Accessed January 18, 2011.

U.S. Department of Energy is a rich, complete site spanning the wide interests of this government agency, from energy science and technology to national security of energy sources. The Department of Energy owns, secures, and manages all the nuclear weapons in the military inventory. Although it concerns all sources of energy, nuclear power is a large component of the mission of the Department of Energy. This site is particularly accessible to students and educators. Available online. URL: http:// www.doe.gov/. Accessed January 18, 2011.

U.S. Nuclear Regulatory Commission is the most comprehensive government site for information concerning the control and safety regulation of the nuclear power industry. It provides a great depth of information concerning nuclear reactors, nuclear materials, radioactive waste, and nuclear security. There is a section that covers public meetings and involvement in nuclear power, as well as a constantly updated event report and news section. Available online. URL: http://www.nrc.gov/. Accessed January 18, 2011.

Index

Note: *Italic* page numbers indicate illustrations.